Architectural
Conservation

For Charles Cockburn

Architectural
Conservation
Principles and Practice

Aylin Orbaşlı

Blackwell
Science

Blackwell Science Ltd, a Blackwell Publishing company
  Editorial Offices:
Blackwell Science Ltd, 9600 Garsington Road, Oxford OX4 2DQ, UK
  Tel: +44 (0) 1865 776868
Blackwell Publishing Inc., 350 Main Street, Malden, MA 02148-5020, USA
  Tel: +1 781 388 8250
Blackwell Science Asia Pty, 550 Swanston Street, Carlton, Victoria 3053, Australia
  Tel: +61 (0)3 8359 1011

First published 2008 by Blackwell Science Ltd

ISBN: 978-0-632-04025-4

Library of Congress Cataloging-in-Publication Data

Orbasli, Aylin, 1968–
   Architectural conservation : principles and practice / Aylin Orbasli.
      p. cm.
   Includes bibliographical references and index.
   ISBN: 978-0-632-04025-4 (alk. paper)
   1. Architecture–Conservation and restoration.   I. Title.
NA105.O73 2008
720.28′ 8–dc22

2007060820

A catalogue record for this title is available from the British Library

Set in 10/12.5pt Sabon
by Aptara Inc., New Delhi, India
Printed and bound in Singapore
by Utopia Press Pte Ltd.

For further information on Blackwell Publishing, visit our website:
www.blackwellpublishing.com/construction

Contents

Preface

Against a backdrop of the uniformity of global architecture and retail empires, the preservation of the built heritage is increasingly being recognised as an important vehicle through which communities can maintain and celebrate their individuality and their diversity. The built heritage is the physical manifestation of a society's history, material evidence of a past way of life, craftsmanship, technique and cultural context. Most significantly it is a finite material and cultural resource which, once destroyed, cannot be retrieved. Architectural conservation plays a vitally important role in ensuring that present and future generations can benefit from the built heritage both in terms of appreciation and enjoyment for its own sake, and for the economic and social advantages that it can bring.

Increasingly, growing concerns for the environment emphasise the importance of maintaining the existing building stock as an alternative to replacement, while at the same time placing new pressures for environmental performance on historic buildings. The body of knowledge relating to the understanding of material conservation is constantly growing, while the meanings attached to cultural heritage are evolving to recognise the value of cultural diversity and context. Some of the conservation methods developed in past decades have now been proven inadequate or inappropriate, while the emerging appreciation of conservation needs for modern materials, such as concrete, is resulting in not only new techniques but also in a re-evaluation of some of the approaches to conservation.

Conservation of the architectural heritage requires inputs from a wide range of professionals, working together as a team. Although many of the attributes required of today's conservation practitioner are based on skills gained in traditional professional training, conservation is a discreet discipline calling on additional skills and values for which a considerable degree of specialisation is needed. A thorough grounding in these specific skills and values is essential if the best interests of the historic building and the sustainability of the built heritage are to be properly served. Moreover, the built heritage concerns not only professionals and craftspeople undertaking repair and conservation work, but also society at large.

In Western Europe, today many construction projects incorporate some component of reuse of historic buildings, and up to a half of the workload of many architectural practices will involve refurbishment of existing buildings, or building within the context of historic town cores and adjoining historic buildings. In other parts of the world too there is a growing recognition that historic buildings have the potential to be reused in imaginative ways.

In architectural conservation no two situations are the same, and each project or challenge calls for a balanced judgement based on sound professional/conservation knowledge and an understanding of the building and its physical and historic context. However, to describe a standard solution for conservation is as impossible as describing a standard methodology for design, as it is at best based on informed and value-based judgement that will vary from case to case.

The concern of this book is the cultural heritage at large and the built heritage in particular. Decisions concerning the conservation of a historic building cannot be isolated from its historic, cultural, social and physical context, its immediate environment, that what it is influenced by and that which it influences. The purpose here is to provide a comprehensive introduction to architectural conservation by placing conservation in its wider physical, social and international context. This book is an introductory text on architectural conservation for students in built environment and related disciplines, for those who own, live in or manage historic properties, those in decision-making positions and those embarking on the field of architectural conservation professionally. The book is not intended as a comprehensive manual of conservation, but as an introductory guide to the principles and practice of architectural conservation today, and as a pointer to other sources of information and training.

Much has been written about building conservation in the past 40 years, both in terms of technical approaches and philosophy. This book builds on this work and knowledge base to provide an easy to access introduction to architectural conservation, and aims to deal with the subject in the context of emerging developments in conservation thinking in the early twenty-first century by drawing on most recent theoretical and technical developments and debates. Whilst it is clearly important to place such discussions in an international context, it is not possible, within the confines of a single volume, to cover all the issues and situations facing architectural conservators worldwide, especially in the sphere of policy and governance. Therefore, this book is primarily focused on experience in the UK, although many of the case studies are drawn from and discussions set in a broader international context since conservation is increasingly informed by international best practice.

In writing this book, I have also drawn on my own research, teaching, practice and consultancy experience in architectural conservation over the past 15 years as well as on discussions with colleagues in practice, at ICOMOS meetings in the UK and internationally and with my students. Amongst many who have provided support and encouragement during the process, I would specially like to thank David Watt for developing the concept and the first outline, Philip Grover for the valuable discussions that enabled the present form and content to develop, Geoffrey Randell for some very useful and constructive comments on an early draft, Michelle Thomas and Stuart Tappin for comments and contributions to the sections on materials and structures, respectively, and Julia Burden and Lucy Alexander at Blackwell for their patience and support. Finally, I am indebted to Simon Woodward for tirelessly reading through the final draft and for his support throughout. Any mistakes and oversights are of course my own.

Part I
Theory and Principles

Chapter 1
Introduction

Most buildings are adapted and added to over time, incorporating layers from different periods and utilising building materials that may have been taken from other, earlier buildings. History is an inseparable part of the environment, living on as physical traces of the distant past. What we see today are not simply 'historic' relics but part of the inherent character of a city, essential cornerstones of its sense of place, to which those living in or visiting it attach meaning and memory. History also resides in the urban form, whether it is the morphology that follows the layout of an earlier period, or the topography that has dictated the historic layout. Rather distinctively, the Piazza Navona in Rome retains the memory of the Roman hippodrome from which it has evolved into a popular city square (Figure 1.1).

In some of the world's rapidly evolving cities, market forces and increasing land values dictate the regular replacement of buildings, in some cases in as little as 20 year intervals. Many contemporary buildings are being constructed for very short lifespans, clearly evident in construction techniques and choice of materials. In the long term, this is not sustainable and at a time of increasing environmental consciousness, old buildings are an important resource that, with sufficient care, can continue to be useful for a very long time. It is often economics that forces a building to become redundant rather than an inability to repair it. At such times, a decision to conserve and reuse requires a level of insight and creativity that is able to overcome the argument for the cheaper replacement option.

WHAT IS ARCHITECTURAL CONSERVATION?

Why conserve?

Historic buildings are an intrinsic part of the built environment. Another reason for conservation arises from a desire to promote national identity or explicitly to stimulate domestic and international tourism. Practically, it makes environmental and economic sense to use what is already there, rather

Figure 1.1   Piazza Navona in Rome retains the form of the Roman hippodrome it has been built over. Remains of the earlier hippodrome can still be traced in some of the basements surrounding the square.

than to waste a resource that remains available for exploitation. There is often a presumption that old buildings present a heavy maintenance burden compared to new buildings, yet all buildings require maintenance, the cost of which is often under-estimated for new buildings.

Some buildings are undoubtedly of national importance and as such will also be respected for the role they play in portraying national identity. For example, the Palace of Westminster in England is not simply valued for its architectural qualities but also for its symbolic role in governance and democracy. In part, heritage can evoke a sense of nostalgia for a past period and is one of the reasons why people choose to visit historic towns and places of historic interest. For the hosts, the economic value and local employment impact of tourism is often a strong reason for conservation. For others, there is added economic or prestige value in owning a historic property, as evident in the choice of historic buildings for company headquarters.

Nonetheless, 'protectionism' continues to draw opposition when it is seen as a barrier to development. The fight for preservation against redevelopment is an ongoing battle, especially with high property prices in inner city areas, such as the City of London. Yet, the listing in 1973 and retention of the old market buildings in London's Covent Garden, supported by the local public in a landmark victory against developers, is proof that historic areas contribute

Figure 1.2 The retention and redevelopment of Covent Garden in London into a major retail and entertainment destination opened the way for other similar structures to be saved and developed in a similar fashion.

not only to the character but also to the economy of cities. Covent Garden today is not only a key destination within London, it has probably increased the value of property in the area more than the proposed office development would have done (Figure 1.2). Furthermore, it has acted as a catalyst for the regeneration of the surrounding area into a major retail and entertainment quarter.

Architectural conservation

In its widest sense, 'heritage' encompasses ruins, archaeological sites, monuments, palaces, castles, vernacular buildings, groups of buildings or ensembles, settlements and urban areas. It also includes the natural heritage, areas of landscape importance and the cultural value of meaning and association. Cultural heritage is formally safeguarded through legislation and its conservation is guided through regulation and management mechanisms established by the legislation. The conservation of each building, place, site or structure will be different in response to the building type, particular situation and use. The approach to conserving extant ruins in a landscape setting will be different from adapting a former industrial building for residential use, yet both will be guided by the same underlying principles. These principles are explained in detail in Chapter 3.

Architectural conservation ranges from preventative maintenance and carrying out minimal repairs to significant modifications, whether part demolition or opening up, to allow a new function to thrive in an existing building. Conservation can involve anything from restoring gilded decorative moulding on the ceiling of a royal palace or remodelling a former factory into a new museum, to maintaining the character of a historic quarter while still allowing it to evolve as a place to live in.

Conservation is the process of managing change while development is the mechanism that delivers change. Historic buildings are not isolated icons; they are part of a larger network of areas, places, towns and landscapes. In making decisions regarding the conservation of the built heritage, the setting and context of a historic building is as important as the building and its material components. Architectural conservation is not simply about buildings, it is also about people, and the approaches to conservation at any time will inevitably be linked to the values of society at that time. The role of the conservation professional is to make balanced judgements that will help maintain the continuity of buildings and townscapes, while serving present-day communities and their needs.

Architectural conservation, like architectural design, is a creative process. Design skills are applied especially in fitting in a new use efficiently and sensitively into an existing building, in accommodating services without them becoming an eyesore and in the careful execution of details, especially where new materials join with existing ones. No two conservation projects are the same, but understanding and then respecting what is already there is a good starting point, recognising a previous conservation technique, designing a new building within the context of a historic townscape. Respecting the existing does not, however, mean that new additions need to blindly copy or mimic the historic precedent (Figure 1.3).

CONSERVATION AS A PROFESSION

The care, protection, conservation and management of historic buildings and sites draws on the skills and know-how of a wide range of professionals, from architects, surveyors and archaeologists to historians, engineers, town planners and landscape architects as well as management professionals. On the one hand, there are the research scientists (materials, environment), on the other, there are the conservators working in the field. Each of these disciplines not only makes a contribution to the field of conservation, but also incorporates some aspect of conservation into the knowledge base of their own professional organisation. There is no one conservation professional, or indeed a single conservation organisation that regulates the field of building conservation. Nationally, on the one side, there are the professional institutions that are fixed and defend the realm of the profession they represent (e.g. Chamber of Architects, Institute of Town Planning) and on the other, there are communities of practice where various professions come together; the

Figure 1.3 Contemporary landscaping in the courtyard is one of the many ways through which this former mill in Prague, in the Czech Republic, has been updated into a modern art museum.

boundaries are often blurred. Internationally, conservation has become an independent discipline, with institutional structures and established knowledge-based communities acting through international charters and doctrines developed by the International Council on Sites and Monuments (ICOMOS) and UNESCO.

Professional roles in building conservation

Conservation is a multi-disciplinary process dependant on teamwork that includes decision makers, a professional team, skilled crafts people and contractors. Each member of the team will bring to the team their own disciplinary knowledge as well as their specialist conservation expertise.

The ICOMOS Guidelines on Education and Training in the Conservation of Monuments, Ensembles and Sites (1993) identify the various areas of knowledge and skills required by conservation professionals (Table 1.1). The key, however, for all professionals is to be able to approach a conservation project with sufficient understanding of the significance of the building or site, its historic context, building technology and techniques and have the necessary knowledge of materials and repair techniques as well as current building

Table 1.1   Skills matrix for the different professions involved in conservation, based on ICOMOS Guidelines for Education and Training in the Conservation of Monuments, Ensembles and Sites (1993) produced by COTAC (Conference on Training in Architectural Conservation).

	a	b	c	d	e	f	g	h	i	j	k	l	m	n
Administrator or owner			x	x				x	x	x		x	x	x
Archaeologist	x	x	x	x				x	x	x	x	x	
Architect	x	x	x	x	x	x	x	x	x	x	x	x	x	x
Art/architectural historian		x	x	x	x	x	x	x				x	x	
Builder or contractor		x		x	x	x	x	x			x	x		x
Conservation or historic buildings officer (Municipality)	x	x	x	x	x	x	x	x	x	x	x	x	x	x
Conservator	x	x	x	x	x	x	x	x	x	x	x	x	x	x
Engineer (Civil or Structural)		x		x	x	x	x		x			x	x	
Environmental engineers			x	x	x	x	x	x			x	x	x	
Landscape architect or historic garden conservators	x	x	x	x	x	x	x	x	x	x	x	x	x	x
Master craftworker		x			x	x	x	x			x	x	x	
Materials scientist		x		x	x	x	x	x			x	x	x	
Building economist (quantity surveyor)				x				x	x	x	x	x	x	x
Surveyors	x	x	x	x	x	x	x	x	x	x	x	x	x	x
Town planner			x	x			x	x	x		x	x	x	x
Curator	x	x	x	x	x	x	x	x	x	x	x	x	x	x

Note: **a** read a monument, ensemble or site and identify its emotional, cultural and use significance; **b** understand the history and technology of monuments, ensembles or sites in order to define their identity, plan for their conservation and interpret the results of this research; **c** understand the setting of a monument, ensemble or site, their contents and surroundings, in relation to other buildings, gardens or landscapes; **d** find and absorb all available sources of information relevant to the monument, ensemble or site being studied; **e** understand and analyse the behaviour of monuments, ensembles and sites as complex systems; **f** diagnose intrinsic and extrinsic causes of decay as a basis for appropriate action; **g** inspect and make reports intelligible to non-specialist readers of monuments, ensembles or sites, illustrated by graphic means such as sketches and photographs; **h** know, understand and apply UNESCO conventions and recommendations, and ICOMOS and other recognised charters, regulations and guidelines; **i** make balanced judgements based on shared ethical principles and accept responsibility for the long-term welfare of cultural heritage; **j** recognise when advice must be sought and define the areas of need of study by different specialists, e.g. wall paintings, sculpture and objects of artistic and historical value, and/or studies of materials and systems; **k** give expert advice on maintenance strategies, management policies and the policy framework for environmental protection and preservation of monuments and their contents and sites; **l** document works executed and make same accessible; **m** work in multi-disciplinary groups using sound methods; **n** be able to work with inhabitants, administrators and planners to resolve conflicts and to develop conservation strategies appropriate to local needs, abilities and resources.

practice and regulations to make informed conservation recommendations and design appropriate solutions.

As in many areas of the building and environment sectors, conservation also has its generalists and specialists. In each of the professions, there will be the generalists who will have some interest in conservation, and there will also be those who have specialised in conservation through training and in the work that they regularly undertake. While in-depth knowledge is the sphere of activity of specialists, they too need to be able to maintain an understanding of the greater whole.

The work of a conservation architect is wide ranging and the projects they are employed on will vary both in size and complexity. The conservation architect is a manager, a designer and often also a report writer. This latter skill is often overlooked in training, but as the following chapters will highlight, at various stages of a conservation project, and increasingly design projects too, the architect is expected to provide convincing written justification and as well as the documents that will become invaluable records for architects and others working on the building in the future.

Conservation professionals may be employed by private and public sector bodies. In some countries, works on historic monuments are carried out exclusively by professionals employed as civil servants in the State protection authorities. In France, a conservation architect can be appointed to a historic building or monument for much of their career, thus building up considerable knowledge and understanding of the historic structure. This is similar to the appointment of accredited conservation architects as surveyor to the fabric to historic churches and cathedrals in England.

Planners also have a wider remit of making decisions concerning the development and management of land and can play an important role in the historic environment and in determining the future of historic buildings. While town planners may be involved in the preparation of master plans and development proposals that will include or impact on areas of historic significance, within local authorities they also have an important decision-making role in determining planning applications concerning historic buildings (see Chapter 4).

Specialist architectural conservators, such as wall painting conservators, not only diagnose the problem and recommend the remedy, they also undertake the conservation on site. A conservation architect or engineer, on the other hand, has the responsibility for diagnosing the problem and specifying the works to be undertaken by a craftsperson or contractor whose work they will inspect. A project manager, on the other hand, must also have sufficient understanding of conservation principles and processes to be able to successfully manage the input of all the various professionals and specialists in a conservation project. For a successful completion the project team needs to be aware of the requirements of each specialist area in order to arrange site timing and sequencing.

The owner, user or manager of a property will in their role as client, be one of the core team members in a conservation project and will play an important role in the decisions that are made regarding a historic property. The expertise and experience of clients will vary greatly from large institutional clients such as the National Trust who are themselves conservation professionals and who have well-established guidelines on conservation policies, to home owners who may be involved in a one-off project. In many reuse and regeneration projects, the client may also be a developer whose main interest is in improving profitability. The more informed a client is of conservation ethics and the procedures, the more able they will be able to actively contribute to the process in the best interest of the historic building. Unfortunately, a great deal of work in the construction industry the world over is defined by short-term objectives and maximising profits rather than making a sustainable contribution.

Craft skills

The effective delivery of a conservation project depends on the availability of properly trained and skilled craftspeople to undertake specialist conservation work, as much as it relies on the inputs of the professionals mentioned above. In some parts of the world, traditional building skills continue to be employed in building practices, ensuring there is a sufficient body of skilled craftspeople to undertake repair and conservation work. In the developed world, however, very few skills continue to be practiced, and there is an acute shortage of people trained in traditional building crafts and conservation.

Conservation works usually include three broad areas of craft skill, namely:

- skills related to the conservation of the building fabric (e.g. stonemasonry and stone repair, traditional joinery, metalwork)
- skills related to the conservation of historic interiors (e.g. decorative plastering, use and application of traditional paints and finishes)
- skills related to landscape conservation (e.g. hedgerows, stone walling and hard landscaping)

The craft skills needed for conservation, however, are not necessarily the same as those required for new build; a stonemason, for example, must be trained separately to become a stone conservator. Building and material conservation is a separate and specialised skill that is acquired through additional training. Most importantly, for these much needed craft trades to survive, those skilled crafts people working in conservation need to be respected in society, and their work fairly rewarded in practice.

Studies in building conservation

Alongside a small number of undergraduate courses, training in conservation is invariably undertaken by professionals qualified in a field of the built environment or environmental science through designated postgraduate study or in practice, supported by CPD (continuing professional development) short courses. Some postgraduate courses also cater for those coming into conservation from different backgrounds or later in their careers. Courses also cover different aspects of the historic environment, from building conservation to urban conservation or heritage management. The multi-disciplinary nature of most courses provides opportunities for participants to learn from one another. Courses on architectural conservation were first started by the International Centre for the Study of Conservation (ICCROM) in Rome in 1962. There are today a great number of training courses, but a majority nevertheless are based in Europe.

However, conservation cannot be separated from the mainstream either. It needs to be recognised as part of the ongoing work of practitioners in the fields of architecture, landscape design, surveying, engineering (structural and environmental) and town planning. At the present time training, in the core disciplines of architecture, town planning, building surveying or structural engineering incorporates little or no emphasis on understanding the existing building stock.

Practice plays an important part in gaining and developing conservation skills. Graduates of conservation courses cannot be expected to know everything about conservation, but as long as they know how to apply the theory, then practice will become an important part of the learning process.

Professional accreditation

Most of the key built environment professions are managed at national level by professional institutions with responsibility for determining required competences for entry and regulating the professional conduct of their members. However, specialisations such as conservation are rarely recognised in the existing structures of professional organisations and trade associations. Accreditation specifically looks at skills and ability in conservation practice, and is the process through which professionals demonstrate their competence in an area of specialism.

In England, a shift in major conservation employers, such as English Heritage and the Heritage Lottery Fund in the UK, to appoint only accredited professionals and crafts people on projects has increased the interest in accreditation schemes for conservation specialists within the professional organisations. Accreditation enables professionals to prove to potential clients their expertise and credibility in the field.

Unlike a degree or a driving license, accreditation is not seen as being for life, but is reviewed at regular intervals, generally every five years. Undertaken by peers, the accreditation process has to remain objective. For an up to date list of accredited professionals to be maintained, the cost of the accreditation process and dissemination of the information is passed onto the membership and must therefore represent 'value for money' to the individual members. In the UK, the ICOMOS Guidelines on Education and Training, noted above, has been used as the basis for informing most accreditation criteria.

Working internationally

Cultural heritage represents the individuality of each community and its unique values. As culture increasingly becomes globalised, the past is often what makes a place different. While information technologies allow conservation professionals to come together and communicate more effectively as an international community, there is still the need to recognise that each culture and community will approach the conservation of its cultural heritage within the framework of its own values. There is also a growing awareness that not all cultures treat architectural heritage in the same way, and even within a community there are likely to be a wide range of values attributed to historic buildings.

Today, architecture has become a global business. Many practices are employing staff from different cultures, operating internationally and entering formal and informal partnerships worldwide. The internet makes it possible for drawings to be produced by separate offices in different parts of the world. For conservation too, many tendering procedures happen on an international basis. The responsibility of the conservation professional working in a foreign country is even more onerous, since proposals and specifications will need to be based on an understanding of the local historic, architectural and cultural context, local material properties, climatic and environmental conditions as well as the current social and economic situation and availability of materials and skills.

At the same time, the remit of conservation continues to grow as it becomes more inclusive and covers a much wider range of type and groupings of buildings and, in part, requires a more pragmatic approach or new solutions and approaches. Material and especially investigative technologies are changing conservation practice, yet in practice a balance needs to be maintained between modern construction technology and the retention of traditional craft practices.

STRUCTURE OF THE BOOK

The book is in two parts. The first part, Chapters 1–4, provides the theoretical basis for architectural conservation while the second part, Chapters 5–8, concerns the applications of the theory into practice.

The next chapter, Chapter 2, sets the context of conservation through a brief history that demonstrates how conservation practices and debates since the nineteenth century have shaped the way conservation is practiced today. Over two centuries, conservation has developed into a truly international field of expertise, yet maintains significant regional differences.

Chapter 3 provides the vital theoretical framework in which conservation practice operates. The chapter starts by offering some definitions of conservation terms and lists the wide range of values that can be attributed to the cultural heritage. Cultural heritage is made up of a multiplicity of meanings that will be interpreted differently by different cultures, different groups within a society and in different eras. The chapter outlines the principal theories and internationally recognised principles of conservation today.

Chapter 4 is concerned with policy and legislation in the protection, conservation and management of the cultural heritage. The chapter identifies the various levels at which decisions are taken in cultural heritage protection and the ways in which different categories of architectural heritage are protected and managed.

Chapter 5 is the first chapter in the second, practical, part of the book and takes an overview of architectural conservation as a process from inception to completion and monitoring of a project. The chapter discusses the various steps that will be undertaken when carrying out conservation work, as well as addressing issues like risk assessment, management and maintenance of the historic building stock.

Chapter 6 starts by identifying the common causes of decay for various materials. The remainder of the chapter is in two sections. The first looks at the incorporation of services into historic buildings, including interventions designed to improve their environmental performance, while the second is concerned with the structural repair of historic buildings.

Chapter 7 focuses on materials, their use, decay and specifically the repair techniques and principles that apply to their conservation. The palette is broad and it has not been possible to cover all variations of materials or their different uses globally. The chapter should be seen as an overview and a pointer to more specialist material-specific literature and sources of information. The science of material conservation is continually evolving and the knowledge base being updated as repair techniques are monitored and new approaches trialled.

Chapter 8 is concerned with the broader remit of design, building reuse and regeneration. Design plays an important role in the conservation and continuation of the historic environment. Various aspects of design in the context of historic buildings and townscape are discussed and related to the principles outlined in Chapter 3.

The Conclusion focuses on some of the new challenges that architectural conservation is facing at the start of the twenty-first century. One of the key debates continues to be around the use of the word 'heritage' and its association with the commoditisation of the historic environment and its simplified value as a visitor attraction. At the same time growing environmental concerns and the effect climate change may have on historic buildings is going to change the way society protects and uses its built resources.

A full bibliography is provided at the end of the book, and a list of recommended reading and resources is provided at the end of each chapter.

FURTHER READING AND SOURCES OF INFORMATION

ICOMOS (1993) *Guidelines on Education and Training in the Conservation of Monuments, Ensembles and Sites.*

Web-based sources

ICCROM, Rome: www.iccrom.org
Institute of Historic Building Conservation (UK): www.ihbc.org.uk
The RIBA Register of Architects Accredited in Building Conservation: http://www.
aabc-register.co.uk/

Chapter 2
International and historic context of conservation

Conservation has always had an international dimension, whether it was the interest in historicism in England triggered by the grand tour visits to Italy in the eighteenth and nineteenth centuries or the notion of rebuilding in the name of restoration and historic integrity that developed in parallel across parts of Europe in the nineteenth century. The emergence of international organisations like ICOMOS (International Council on Monuments and Sites) in 1965, followed by UNESCO's launch of the World Heritage Convention in 1972, has firmly established conservation as an international concern. The conservation movement has nonetheless remained anchored in the West and has all too often reflected predominantly Western values. Only more recently has there been the recognition that the relationship of 'culture' to the built form can differ widely in different cultures around the world. At the same time, the remit of cultural heritage has expanded significantly to encompass historic towns and villages, industrial sites, twentieth century and modern movement architecture as well as the physical and social context of the historic environment. This much wider definition and appreciation of the many facets of the built heritage can also be seen in the change in professional terminology from 'historic monument' to 'cultural property', and more recently to the much debated 'cultural heritage'. Architectural conservation today sits as a discipline between the science of materials conservation and the sustainable management of the built heritage.

This chapter sets out a brief narrative of the history of architectural conservation from the late eighteenth century through to the early twenty-first century. The purpose is to highlight and explain some of the influences that have shaped the principles and ethics that provide the cornerstones of conservation today. It will also chart some significant shifts in thinking on what constitutes cultural heritage, what should be protected and how this should be approached.

Figure 2.1   The church of San Lorenzo in Miranda in Rome, Italy, is built into the former Roman temple of Antoninus and Faustina.

HISTORIC CONTEXT

The symbolic and nostalgic connotations of historic monuments

Throughout history, monuments have been regarded as being 'symbolic' and their preservation and maintenance in later generations has tended to be seen as an obligation. The practice of restoration and repair of monuments can be traced back to ancient times. The coming of Christianity, for example, saw many Roman temples and basilicas adapted for use as churches (Figure 2.1). In other instances, the ruins of older buildings were utilised as convenient quarries of building materials. The Venetians also prolifically reused building materials and parts of buildings so as to save on the costly transport of new materials.

A more conscious recognition of historic buildings as 'heritage' assets is seen during the Renaissance when an interest in monuments from the classical period resulted in their restoration. An established theory of conservation first developed in Italy, UK and France during the eighteenth and nineteenth centuries, and later influenced other countries including the US and Canada. In this period, the theory was very much applied to the restoration of monuments.

The eighteenth century 'grand tour', part of a young gentleman's education, not only popularised collecting antiquities but also generated an interest in the protection of medieval monuments back in Britain. Also, at this time historic monuments, and ruins in particular, were being placed in the context of the picturesque. Ruins, either real or fabricated, were regularly incorporated into the landscape schemes of the time, while engravings in the picturesque tradition of exotic buildings and ruins emphasised what remains a European fascination with the old and dilapidated.

'Restoration' in the nineteenth century

By the nineteenth century in England, Germany and France, the word restoration had become synonymous with the reordering and reconstruction of monuments, often with little proven evidence, to what was thought to be the original design intention or simply to establish an assumed symmetry. The reinterpretation of mainly Gothic churches, in the name of purity of style, paid little respect to authenticity or indeed architectural evidence. In many cases, layers of later additions were removed from buildings to achieve stylistic consistencies including new additions that were deemed to be in keeping with the desired style.

In France, this period of intensive restoration is commonly associated with the name of Viollet-le-Duc, who exemplified his interest in the medieval through his perfectionist restoration works. There were often stylistic transformations in the name of unity and based on his interpretation of the architecture of the period and the 'original' design concept. A similar desire for unity in style is seen in Sir Gilbert Scott's work on English churches.

At the end of the nineteenth century, clearing areas around important monuments and displaying them in the midst of large parks and gardens was a popular approach to their preservation and presentation. Haussmann's reordering of Paris not only swept grand boulevards through dense urban areas but also cleared much of the cluster of urban fabric from around the Cathedral of Notre Dame. The cathedral, like many other European cathedrals of the time, would originally have been conceived to emerge over rooftops and be glimpsed from side streets. This setting was, however, lost in Haussmann's planning (Figure 2.2).

The search for authenticity

By the latter half of the nineteenth century, a growing 'anti-restoration' movement started to emerge as an opposition to some of these practices. Despite the growth of most modern day conservation principles deriving from these opposition movements, the question of rebuilding continues to be debated to

Figure 2.2   The forecourt and the wide roads on either side make the Cathedral of Notre Dame a recognisable and manageable tourist destination, but the original design intention of rising from a dense urban fabric has been lost.

this day. Proposals to rebuild monuments or to restore a building to a specific time in its history, including the addition of features that 'may have been used' at the time, continue to be proposed. In Germany, proposals have been made to rebuild the demolished royal palace in the heart of Berlin, following demolition of the East German Palast der Republik (People's Palace) on the site. The argument is that the former 'Prussian' palace building, about which there is limited detailed documentary evidence, would be a better reflection of the German identity and would allow for the convenient removal of a relic of the former German Democratic Republic. Other arguments are that a 'historic' palace is a far better fit for the character of the famous Unter der Linden Avenue than a mid-twentieth century structure of the communist era.

Opposition to the ongoing restoration practices were consolidated in England through the creation in 1877 of the Society for the Protection of Ancient Buildings (SPAB), with William Morris as its honorary secretary. Morris argued that to restore and to copy destroyed authenticity. He saw protection as doing no more than necessary to keep an ancient building in sound condition. Morris's manifesto has become a template for modern conservation policy, and SPAB continue to be a highly regarded conservation advisory body in the UK.

John Ruskin, another prominent SPAB founder member, suggested that historic buildings needed to be maintained within their setting rather than being isolated in a landscaped park. He became an advocate of repairs rather than stylistic replacements, and of honesty with any intervention, avoiding decorative carving on stone replacements and dating all new work. It was Ruskin's criticisms of restoration practices of the time that led to the replacement in England of the word 'restoration' with 'conservation'. He associated the marks of age on a building or work of art as part of its beauty and acquired character. According to Ruskin, real heritage lay in the genuine monument, not in modern replicas. Ruskin also pointed out the value of historic cities, not only in terms of single monuments, but in the collective value of buildings, streets and spaces that made up the character of old towns, which he feared were being lost to modern developments and street widening schemes.

The SPAB Manifesto written by William Morris and other founder members in 1877, introduced the 'conservative repair' philosophy. The manifesto continues to form the philosophical basis for the Society's work to this day and inform conservation practice. Morris and his colleagues were not alone and mid-nineteenth century restorations of antiquities in Rome can still be seen with new sections in brick to distinguish them from the original marble (Figure 2.3). While Ruskin and Morris developed the theory it was the architect Phillip Webb who was implementing their principles, and the one who recognised that theory and application were not always compatible.

The first British Ancient Monuments Act was passed in 1882. More than a decade later, in 1895, the establishment of the National Trust also played an important role in saving and preserving a significant number of historic buildings. The National Trust was a product of private enterprise not seen before, purchasing or obtaining as gifts areas of natural beauty, and buildings or groups of buildings of historic importance. From the outset, the National Trust was consulting SPAB for the preservation of these assets it was accumulating for the nation. The Trust remains one of the foremost conservation organisations in England today and has been a model in the formation of a number of similar trusts worldwide. Much of the principles of conservation adapted today and manifested in the twentieth century have their origin in the debates and practices of the nineteenth century.

Figure 2.3   Restoration work in Rome, Italy, clearly distinguishing new material by using brick in contrast to the marble original.

CONSERVATION IN THE TWENTIETH CENTURY

Built heritage as a symbol of national identity

The growing interest in heritage in the twentieth century, combined with nationalistic feelings in the aftermath of two World Wars and the economic value associated with cultural tourism, has defined conservation in Europe in the latter part of the twentieth century. Conservation theories continued to be discussed and new conservation technologies developed, most notably in Italy. The destruction caused by the wars, followed by the emergence of new nation states and the shift of power from major colonial empires resulted in new meanings and associations being attributed to the architectural heritage. During this time, the conservation of buildings and places that were associated with a national or ethnic identity often gained priority.

After the war, it was important for the dignity of nations to rebuild national monuments that had been lost or severely damaged. However, in many places and cases, the principles of conservation gave way to rebuilding. The precedent was set in Ypres, where options for a new replacement city or a ruin were abandoned in favour of completely rebuilding the town to its appearance before the war, specifically for its symbolic meaning. As a result, many buildings in Ypres were built from scratch, sometimes merely on photographic evidence. In other places, the emphasis remained on places of public importance such as monuments or major squares that were easily identifiable and therefore symbolic.

The cycle of destruction continued and in the Second World War many European cities, especially in Germany and Eastern Europe, were once again severely damaged. In the aftermath, a similar approach to rebuilding was favoured as the symbolic significance of the architecture was of greater national importance than strict adherence to conservation theory. The historic centre of Warsaw was completely rebuilt on the basis of extensive pre-war documentation. More significantly, the 'historic centre' of Warsaw has since been inscribed on the UNESCO World Heritage List. In Germany, facades looking onto major squares were recreated as facsimiles of the previous buildings, while the buildings behind them were conceived as more contemporary spaces. In other places, modern solutions were favoured in the rebuilding, the notable examples being Coventry Cathedral in England and Cherbourg in France where a large part of the rebuilding followed 'modernist' principles and the use of precast concrete.

The Convention for the Protection of Cultural Property in the Event of Armed Conflict, also known as the Hague Convention, became the first UNESCO Convention concerning the cultural heritage when it was launched in 1954, although debates on the issue dated back to 1907 (see Table 2.2). The convention explicitly calls on its signatories to respect cultural heritage in 'enemy' lands in times of armed conflict. However, the principles behind the Hague Convention, and later agreements that cultural property should not be damaged has yet to prevent extensive destruction of cultural property during times of conflict.

Cultural heritage is undoubtedly a symbol of national and cultural identity. As international and civil warfare continues to damage and destroy these symbols, they will also continue to be rebuilt. The historic centres of the former Yugoslav cities of Dubrovnik, Sarajevo and Mostar's famous bridge have all been rebuilt following conflicts in the 1990s. But it is for the very reason that they symbolise national identity that the historic centre of Dubrovnik, the library of Sarajevo and the bridge of Mostar were targeted in the first place.

International conservation charters

A meeting of mostly European representatives, held in Athens in 1931 to discuss the conservation of architectural monuments, produced a series of recommendations. Later to be known as the Athens Charter, the recommendations formed the first international document outlining modern conservation policy. The charter discouraged stylistic restoration in favour of conservation and repair that respected the various changes a building would have gone through. Issues such as community, maintaining monuments in situ and in character of their setting were discussed, as were the application of modern technologies in conservation practice.

In 1964, representatives of 61 countries came together in Venice to revise the 1931 Charter. The meeting came at a critical time, as a reaction to the stylistic restoration that had taken place after the war and to the modern movement urbanism that was taking over city centres with little consideration of the historic environment. The outcome, the Venice Charter, is a key turning point and highlights issues that were being debated at the time and which led on to shape conservation in the latter half of the twentieth century. It moved away from the idea of the individual monument, towards defining context, and most significantly set out some sound principles for conservation and extending the understanding of historic monument from individual building to also incorporate areas, both urban and rural. Although countries around the world have signed up to the Venice Charter, it is not a legally binding document, but one of ethics that can guide the formulation of national policy and inform practice. Like all documents of its kind, the charter is open to varying interpretations.

Much has been written about the Venice Charter since 1964, its various anniversaries celebrated and numerous attempts made to revise it. One of the initial objectives of the charter was that it should be seen as a general framework for each nation or culture to develop a more customised charter that suited their own heritage needs. This has been the case with the Australian Burra Charter that is also regularly updated and a more recent 'Chinese Principles'. Originally drafted in 1979, the Burra Charter has been revised periodically, most recently in 1999. Building on the Venice Charter, it has brought greater clarity to a number of issues as well as definitions that have become universally accepted, including the introduction of 'the concept of place'. It is also a reflection of Australian concerns for the conservation of up to 40,000 years of indigenous heritage integrated with nature and based on oral traditions and overlaid with little more than 200 years of the European style heritage of the settlers.

The Venice Charter must be seen as an embodiment of European values and concerns of the post-war period. Its adaptation to North America, Australia, South America or Asia needs to consider a certain evolution. Nonetheless, the framework of the Venice Charter has single-handedly influenced conservation policy worldwide. As Table 2.1 lists, the Venice Charter has also led to a wide range of more detailed and subject-specific charters and guidelines. To some, these are often seen as being too general and therefore open to different interpretations in practice. Charters drafted in one language do not always translate very well into other languages, or to other cultures and their values. Furthermore, whether a charter can fully embody international cooperation is of course debatable, as is the relevance of a set of international guidelines that will be acceptable and applicable the world over. As they stand, the charters are merely advisory documents on ethics and principles of conservation.

Table 2.1   ICOMOS charters and international guidelines.

ICOMOS Charters
International Charter for the Conservation and Restoration of Monuments and Sites
  (The Venice Charter) – 1964
The Florence Charter on Historic Gardens and Landscapes – 1982
Charter for the Conservation of Historic Towns and Urban Areas (The Washington
  Charter) – 1987
Charter for the Protection and Management of the Archaeological Heritage (The
  Lausanne Charter) – 1990
Charter for the Protection and Management of the Underwater Cultural Heritage –
  1996
International Charter on Cultural Tourism – 1999
Principles for the Preservation of Historic Timber Structures – 1999
Charter on the Built Vernacular Heritage – 1999
ICOMOS Charter – principles for the analysis, conservation and structural
  restoration of architectural heritage – 2003
ICOMOS Principles for the Preservation and Conservation-Restoration of Wall
  Paintings – 2003
Guidelines
Guidelines for Education and Training for the Conservation of Monuments, Sites
  and Ensembles – 1993
The Nara Document on Authenticity – 1994

In 1965, the signatories of the Venice Charter established ICOMOS as an international non-governmental organisation. The Venice Charter became the chief doctrinal document for ICOMOS on its foundation, and the international conservation movement was established. Today, ICOMOS has a membership of over 120 countries with a secretariat in Paris and is a recognised adviser to UNESCO on cultural heritage sites. A series of charters responding in more detail to specific conservation issues or to regional differentiations have been developed by ICOMOS's extensive network of International Scientific Committees. These committees make ICOMOS a scientific organisation bringing together international expertise, setting an international agenda for conservation and, more importantly, establishing international benchmarks through meetings, conferences and international charters.

The non-Western perspective

UNESCO's 1995 Nara Document on Authenticity states that: '*... the cultural heritage of each is the cultural heritage of all. Responsibility for the cultural heritage and the management of it belongs, in the first place, to the cultural community that has generated it, and subsequently, to that which cares for it*' (Nara Document on Authenticity, 1995).

Signed in Nara, Japan, one of the objectives of the document is to recognise the different associations' different cultures have with the cultural heritage

Figure 2.4   In Japan, like in other South East Asian countries, dismantling and rebuilding buildings of cultural significance is seen as valid means of conservation.

and the concept of authenticity, while also ensuring an understanding of a common heritage of mankind.

While the modernisation of the Western world since the Enlightenment took on a model of separating religion and the spiritual from the scientific and 'tangible' (rational), this has not always been the case in other parts of the world. Furthermore, different cultures have different relationships with their natural and built environments. They have different realities, customs and beliefs when it comes to the conservation of the cultural heritage. The idea of renewal common to Eastern cultures is fundamentally contradictory to the principles of maintaining original fabric (Figure 2.4). In Africa, cultural heritage resides in oral histories that traditionally were passed down through generations, and that are rapidly dying out today. In other places, economic or political factors are a major hindrance to conservation.

In Japan, for example, dismantling and rebuilding and repair is common practice, partly related to Shinto traditions when temples and shrines were temporary structures built for special occasions. Wealth of historic association is an important aspect of protection, and value attached to intangible cultural properties incorporating performing arts and applied arts. Conservation policies imported from the West in the late nineteenth century have been incorporated and adapted into Japan's cultural and natural conditions. In Thailand, reconstruction is seen as traditional Buddhist practice and the Venice Charter has been adopted to respect the spiritual values of a place and the greater extent of renewal that takes place in local traditional restoration practice through the 1985 Bangkok Charter. Local charters need to reflect local values and recognise the socio-economic framework in which conservation will take place.

A number of well established and recognised international organisations play an important role in influencing conservation internationally through example, research, doctrine, training, awards and financing. Alongside ICOMOS, its sister organisation ICOM (International Council of Museums) is involved in the conservation of museum objects, ICCROM (The International Centre for the Study of the Preservation and Restoration of Cultural Property) plays an important role in conservation training worldwide. UNESCO is the arm of the United Nations that is involved with culture, including World Heritage Sites. In Europe, the Council of Europe, a political body, made up of 46 state members, includes the promotion of know-how on culture and heritage as part of its remit. The non-political Europa Nostra is a pan-European federation for cultural heritage and an advocate of high standards in conservation including an award scheme. The Aga Khan Trust for Culture is positioned in the Islamic world, where through various programmes it directly implements conservation projects alongside a prestigious award scheme. The US-based Getty Conservation Institute is a highly acclaimed centre for conservation research, publication and grant funding. The World Monuments Fund is a private, not-for-profit organisation mainly concerned with endangered cultural sites.

The urban conservation movement

By the 1950s and 1960s, many of Europe's historic centres had become run down areas, often seen as an obstacle to new transport developments, city centre renewal and commercial development schemes. Accelerated by increasing property values, historic buildings and entire neighbourhoods were demolished in the name of development and progress. These 'regeneration' schemes of the time paid little heed to the original layout or character of these areas and wholesale replacement of historic cores in a completely new layout and scale was common practice. The advocates of the modern movement took little interest in historic buildings or historic character in their search for new utopian environments. In the event, it was a civic movement and pressure from amenity groups and a public outcry against the loss of heritage that started to bring about change. Public voice and demonstrations started to save some buildings from demolition in the 1970s. Today, organisations like Save Britain's Heritage and Save Europe's Heritage, amongst others, continue to campaign against the destruction of historic buildings and the erosion of their settings.

Although the first legislation in conservation appeared in the nineteenth century, it was only in the 1960s that comprehensive, area-based conservation legislation emerged, allowing for the first time the designation of historic urban areas to ensure their protection and conservation. In England, the Civic Trust, established in 1957, brought together local amenity societies to establish a joint voice on townscape, including an award scheme that encouraged new build and conservation schemes in historic urban areas. This development was followed by the 1968 Town and Country Planning Act and the growing reactions against comprehensive redevelopment schemes.

The Council of Europe's European Architectural Heritage year initiative in 1975 played an important role in raising awareness of the value of the built heritage to towns and cities, encouraging civic authorities to tackle some of the problems facing them. The objectives of the year were to awaken a greater interest in architectural heritage and the conservation of the character of old towns and villages by assuring a living role for historic buildings in contemporary society. Local authorities across Europe were asked to prepare for the year and grants were made available for a wide range of environmental improvements. By the end of the year, conservation had been brought into the context of urban and regional planning. Today, most historic towns and centres in Europe are designated protection and conservation areas and many have become lucrative tourism destinations, a legacy in part at least of this Council of Europe initiative.

The period of rapid technological, social and urban change that marked the twentieth century, also brought about pressures for, and an interest in conservation in the West. Similar developments in other parts of the world in more recent times have not, however, necessarily been met with a greater public demand for conservation. For a start, in most instances the problems faced have been greater, the conditions within the historic centres more dire, the economic situation poorer and the pressure for renewal and development much greater. Area-based conservation and the designation of conservation areas, while now a common component of conservation law and policy in West, remains in its infancy in many parts of the world.

UNESCO and the World Heritage Convention

The Convention Concerning the Protection of the World Cultural and Natural Heritage, also known as the World Heritage Convention, was adopted by UNESCO in 1972 and ratified by member States in 1997 (see Table 2.2). Through the convention, UNESCO invites cultural, natural and mixed sites

Table 2.2   UNESCO conventions relating to cultural heritage.

Convention for the Protection of Cultural Property in the Event of Armed Conflict – The Hague Convention – 1954

Convention on the Means of Prohibiting and Preventing the Illicit Import, Export and Transfer of Ownership of Cultural Property – 1970

The Convention Concerning the Protection of the World Cultural and Natural Heritage – 1972

Convention on the Protection of the Underwater Heritage – 2001

Convention for the Safeguarding of Intangible Heritage – 2003

Figure 2.5   The World Heritage Emblem can be displayed at sites inscribed on the World Heritage List. The central square represents form created by man and the circle symbolises nature, the world and protection.

of universal significance to be nominated for inclusion on the World Heritage List (Figure 2.5).

Nominations for inscription on the World Heritage List are made by a State Party and evaluated by the World Heritage Committee, which is made up of elected members. The convention is managed by the World Heritage Committee and the permanent staff at the World Heritage Centre in Paris. Funding for the centre is linked to UNESCO membership and annual dues from member states. Operational Guidelines for the Implementation of the World Heritage Convention are set out to guide applicants on the principles of the Convention. The Guidelines were first drafted in 1977 but are constantly reviewed and updated and form the basis for any application and its evaluation. The current Operational Guidelines not only require a management plan to be in place, but seek assurance that the management plan has reached consensus with all stakeholders including the local community. This is a notable shift from the intellectual and elitist attitudes present in the early years of the convention.

On the thirtieth anniversary of the convention in 2002, there were a total of 563 sites in 125 countries inscribed on the list and by 2006, the number of World Heritage Sites had increased to more than 800. Most notable is the significant change of scope over the three decades, with more recent nominations including a modern movement university campus, a concentration camp and a system of canals. Furthermore, the cultural element, at times intangible, of natural sites and the importance of setting for cultural sites, has also now

Figure 2.6   Uluru Kata Tjuta National Park in Australia, originally inscribed on the World Heritage List as a natural site in 1987 and re-nominated as an associative cultural landscape in 1994 in recognition of the significance of the landscape to the livelihood and beliefs of the aboriginal peoples. (Photograph by John La Salle.)

been recognised (Figure 2.6). The World Heritage List in this respect might be seen as a kind of measure of heritage understanding, reflecting changes in the international community's view on what constitutes heritage and how inclusive it is.

SHIFTING EMPHASES

As demonstrated by the growing and increasingly diverse World Heritage List, the remit of cultural heritage continued to grow towards the end of the twentieth century. Purely aesthetic and historic considerations that governed decisions concerning cultural heritage earlier in the century started to give way to a broader understanding of cultural values. For instance, the technical, engineering, aesthetic and social values of industrial heritage have been recognised while buildings built in the twentieth century are gradually being listed and conserved and taking their place in the 'past'. When ICOMOS was established in 1965, the concern of its scientific committees, of which there were initially five, were mainly monuments and conservation techniques relating to structures and materials. The current 28 scientific committees, cover diverse areas of interest from underwater archaeology to cultural routes and the Polar heritage.

Conservation-led regeneration

By the 1990s the picturesque historic towns of Europe and other parts of the world had become well established and popular tourist destinations. For some, tourism and the associated service economy had become the major industry on which the economy depended. The unprecedented success of recognised destinations subsequently led other, less well-known historic towns and cities with historic quarters to also promote themselves as cultural tourism destinations. Tourism is an important means through which architectural heritage contributes to city economics and there has been an increasing realisation that cultural heritage can be a vehicle rather than a hindrance to urban regeneration.

Whereas in the 1970s, much of the focus of urban conservation was on picturesque historic towns, by the 1980s the local authorities of larger and former industrial cities were also identifying suitable urban renewal and regeneration projects for their own historic quarters. Since then urban regeneration projects have revitalised derelict waterfronts, harbours and ports, creating new entertainment or residential quarters, although more usually the choice has been for mixed-use developments. The contribution of the built heritage has thus become an accepted component of urban regeneration. Increasingly more urban renewal programmes are focusing on the existing building stock as a valuable commodity that creates part of the character that defines a place. Nonetheless, it is only relatively recently, for example, that World Bank funding programmes have included cultural heritage within their social and economic development remit.

The industrial heritage

Many urban regeneration projects have highlighted the value of industrial heritage, as former quarters of mass production and old dockyards have been transformed into new and vibrant urban districts. However, the conservation of the industrial heritage has brought new challenges with it. Firstly, it consists of buildings and structures that are not necessarily deemed to be 'pretty' or attractive in the manner that other historic buildings are. Secondly, it represents the legacy of a period of history that is simultaneously associated with progress and innovation as well as with hardship and suffering. Thirdly, a significant shift has taken place, from working industrial places of machinery, noise and dirt to the sanitised uses of today. Maintaining machinery can be a major challenge and often land around industrial zones can be contaminated.

While in larger city centres former industrial areas tend to evolve into 'hip' living quarters, such as the Canal Street Area in Manchester, UK, or New York's SoHo district, there is less viable use for the large number of industrial buildings in more isolated locations such as the Yorkshire Pennines, where the decline of industry has also resulted in population decline. Even in a relatively large city like Liverpool, the vast area of industrial buildings

Figure 2.7   Liverpool's Stanley Dock where the scale of some of the buildings is a major challenge for reuse and regeneration of the area.

around Stanley Dock, now part of the Liverpool World Heritage Site, remain derelict and are largely unused (Figure 2.7).

The industrial heritage also presents a new scale of heritage monument that has to be dealt with. Not only industrial cities, but also the scale of these vast areas of industrial activity, were unprecedented for their time. In redundancy, the same capacity is also in search of a new function and ways of maintaining the significance and integrity of the heritage. For example, the historic dockyard at Chatham on the Thames Estuary has become a major challenge for regeneration agencies and those concerned with the conservation of its historically significant buildings. Although over 30 buildings have been restored, some of which are open to the public as a visitor attraction, it has not been possible to find a feasible new use for all the large sheds, especially given the high costs associated with their conservation, including the structural works that need to be undertaken. A dockyard, like any other industrial complex, includes elements such as rail tracks and cranes associated with its former industrial use. Stripping places from these now redundant features also affects the integrity and inherent character of the place, as well as causing the loss of vital historic information. Addressing the legitimate conservation argument for their retention has to be weighed against the interest of public safety and the availability of sufficient funds for maintenance.

Another similar legacy is emerging in the relics of twentieth century military heritage, not only of the two World Wars but also of the Cold War that followed. Large-scale military bases that are now being decommissioned are significant testimonials to a period of history and changing warfare. The architecture of warfare has always been part of the cultural heritage, but although medieval castles have become favoured visitor attractions, this is yet to be the case for most concrete bunkers, although several Cold War command and control bunkers in the UK are now being opened to the public as a visitor attraction. The remnants of more recent wars take on not only new material characteristics but also new typologies. The use of highly reinforced concrete in the construction of many of the buildings from this period, make them equally difficult to eradicate or to conserve. Similarly, will the great power stations or petroleum refineries of today become the heritage concern of tomorrow?

The legacy of the twentieth century

The designation and listing of twentieth century architecture is also relatively recent, especially those related to the modern movement and more recent periods. Few countries today have legislation in place to cover buildings from any part of the twentieth century. Compared to established norms for protecting older buildings of historic and architectural interest, the listing and protection of more recent buildings in the face of continuously shifting value judgements is often keenly debated. The first issue will be when a building is worthy of protection. In England, a '30-year rule' is applied. A building seen as a utopian ideal for living and then as a blighted tower block may be viewed as an architectural icon within the course of 30–40 years. Where as in earlier periods, creative genius was focused on works of outstanding importance and therefore major monuments, in the post-industrial era of mass production, the objective of many modern movement protagonists was the production of 'designed' items for the less affluent classes and the mass market. There is, therefore, a much greater quantity of 'designed' objects and structures of special architectural and often social historical interest.

There are other practical issues to consider in the conservation management of buildings that are still very much in use and still changing and evolving, whether they are company headquarters or residential flats. Many buildings, especially in the 1950s and 1960s, were experimental in design, structure and materials that may not have proven successful in terms of technical or social performance. Substandard buildings of earlier periods have simply not survived, but the same is not the case for more recent buildings. Technical issues of dealing with the conservation of new building technologies, including 'revolutionary' construction systems and new materials for which the most appropriate repair techniques have yet to be developed, are just some of the issues facing the conservation of modern movement buildings (Figure 2.8). In some cases, established repair philosophies are also challenged, such as the reversibility of a concrete repair or patch repairs to a modernist building that

Figure 2.8 The conservation of modern movement buildings, such as the Poimio Sanatorium by Avar Aalto in Finland, have presented conservation professionals with a number of new issues concerning materials as well as philosophies to be adopted.

was conceived as a flawless surface. The impossibility or high cost of manufacturing replacements for components that are no longer manufactured but that are integral to the design is another problem that is frequently encountered. Moreover, some materials have proved in appropriate for their purpose, whilst others like asbestos have since proven to be hazardous.

One positive factor in the argument for conserving modern buildings is the fact that there is often considerable information available on the original design intention of relatively recent buildings, in the form of preserved records, published reviews and indeed from the architect who may still be alive. The principles for the conservation of modern buildings, especially in relation to accepted conservation philosophies and restoration are still open to debate.

Cultural landscapes

Cultural landscapes are an outcome of a culture that has created them, through beliefs, ideas or physical interventions, and denote the relationship between human beings and their environment. The increasing recognition in conservation circles of cultural landscapes is a move away from separating buildings from their environment and reflects the understanding of cultural heritage as being an entirety. Nature and culture cannot be separated and they often form an integral whole, each informed by the other. The landscape is not simply a visual backdrop to the built heritage.

In a cultural landscape, each historic building is contextualised in a series of relationships to other buildings with which it belongs, whether it is a multi-layered townscape or other buildings of the same period, designed by the same architect or built from the same materials. A historic railway constitutes railway buildings such as stations, railway lines and engineered structures such as bridges, the engines and mechanics that make it work, the landscape through which the railway passes, settlements along the route and the route itself. The French World Heritage Site network of canals, Canal du Midi, is to a large extent a landscape that has been transformed by the very presence of the canal. A historic city is not an isolated entity either, but linked to a greater network of towns and cities whether through alliances, competition or trade. However, the context, should not only be taken as townscape, landscape or setting but also the social and political setting in which they are placed. Today, the archaeological site of Çatalhöyük in central Turkey is viewed as a singular prehistoric site, the earliest known settlement of its size in Anatolian prehistory. Yet, Çatalhöyük was part of a network of farming and trading communities that defined the Anatolian landscape of the time, and part of its cultural heritage value must be seen in terms of this larger, but also significantly altered landscape.

The World Heritage List has again been influential in defining the concept of a cultural landscape and a significant number are now listed as World Heritage Sites. The protection and management of such entities that are continuously changing is a major challenge. Compared to the more static qualities of the built heritage, the landscape is an ecosystem in constant evolution. Despite repeated references to the importance of context and setting for historic buildings since the nineteenth century, legislation the world over continues to operate mainly within the boundaries of listed monuments and buildings, to which protection is afforded and rarely within the broader landscape setting.

Intangible heritage

In some cultures, spiritual value or place value may be more important than the value or authenticity attributed to physical remains or evidence of the past. Not all aspects of cultural heritage are physical by nature, and most cultural heritage sites also include other intrinsic values. All places of worship, for example, also behold a strong sense of spirituality. There are also examples where the spiritual value of the place is contested amongst different religious groups, such as the Temple Mount in Jerusalem or the Hagia Sofia in Istanbul,

Case study: Cappadocia National Park cultural landscape

The Cappadocia National Park in Turkey is listed as a World Heritage Site for its natural rock formations as well as the man-made interventions within this landscape. Early Christians fleeing persecution not only hid in a series of underground settlements dug into soft tufa stone, but also created some spectacular churches that they carved into rock and decorated with elaborate frescoes. As well as being a convenient material to carve out of, the tufa is also prone to rapid erosion. A number of churches have already been lost, and conservation scientists continue to seek solutions to halt the process on others. It is, however, a case of man against nature, and if the Cappadocia Valley is viewed as an evolving cultural landscape, then nature is following its own natural evolution, albeit at the expense of the man-made cultural heritage (Figure 2.9).

Figure 2.9   Cappadocia National Park in Turkey where the naturally eroding rockface exposing some of the medieval churches carved into the rock.

originally built as a church during the Byzantine period, then used as a mosque by the Ottomans and now open to the public as a museum.

The intangible heritage, as defined by the 2003 UNESCO Convention for the Safeguarding of Intangible Heritage, is '*the practices, representations, expressions, knowledge, skills – as well as the instruments, objects, artefacts and cultural spaces associated therewith – that communities, groups and, in some cases, individuals recognize as part of their cultural heritage*,' and includes dance, music, endangered languages, drama, food cultures, crafts and other traditions with the aim of safeguarding them against change. For the built heritage, intangible heritage refers to the values that are not apparent or related to the physical fabric of the building, but are an integral part of its significance. This may be associations with certain personalities or events in history, personal memory or spiritual values associated with buildings and places. A church, for example, needs to be recognised as the sum of its architectural components embodied in its spiritual meaning, which may be different for each of its users. The intangible heritage also includes craft traditions and methods, the preservation of the knowledge that created the buildings.

Integrated approaches

With expanding knowledge, the way we are looking at historic buildings is changing. There is a better understanding of a much wider range of values and historic value itself may not always be the most overriding. At the same time, context and setting are much better recognised as part of the integral value of cultural heritage. Physically, this means a much wider remit for conservation and a growing number of objects and places that are considered to be worthy of protection. Current approaches to conservation advocate much greater integration between disciplines, buildings and their physical environment and social context. Understanding the intangible values of heritage also highlights the importance of multi-disciplinary approaches discussed in Chapter 1, including closer working relations with social science disciplines such as anthropologists.

The protection of sites and monuments has also brought about their enclosure and other means of discouragement, adding a sense of exclusivity. As the remit of cultural heritage has widened, so has its impact on the daily lives of communities. While some buildings and monuments are proclaimed as symbols of national identity, others signal to individuals a sense of belonging or an association with a place or community. Decisions relating to the conservation of the cultural heritage, from monuments to neighbourhood renewal, cannot be isolated from the local communities who will in many cases be their best guardian.

Starting with urban regeneration projects, there is an increasing realisation that conservation is a local concern and that local residents are vital partners in conservation. Historic buildings are now expected to be more intellectually, socially and physically accessible to their communities. Accessibility is not simply a physical issue, it is about creating a better environment and better understanding of the cultural heritage for all. Historic buildings can be better appreciated through positive experiences. Adapting historic buildings to provide users with improved physical access is discussed in more detail in Chapter 5.

SUMMARY AND CONCLUSION

Whereas the conservation movement started out as being scholarly and elitist it has developed into a more popular and inclusive movement and a strong political tool for successive governments. Conservation has, however, always embodied within it a series of conflicts, whether of rebuilding to an aesthetic ideal or carrying out minimal and honest repairs; of ownership and custodianship; differing cultural values relating to the tangible and intangible dimensions of the cultural heritage, and social and ethic concerns.

There will always be opposition to conservation and each new movement or widening of scope is often met with cries that 'too much is being conserved'. Yet, the rate of destruction of the cultural heritage the world over is considerably higher than the level of protection that can be offered. Even in England, only 2% of all buildings have legal listed protection, hardly a significant figure considering the value of the built heritage to the character of places and social identity. Conservation has to be seen as more than simply protecting historic places and buildings, but as a process that enables them to be maintained and changed if necessary but always recognising the values that these heritage assets stand for.

FURTHER READING AND SOURCES OF INFORMATION

Erder, C. (1986) *Our Architectural Heritage: From Consciousness to Conservation.* Paris, UNESCO.

Hunter, M. (ed) (1996) *Preserving the Past: The Rise of Heritage in Modern Britain.* Gloucester, Great Britain, Alan Sutton Publishing.

Jokilehto, J. (1999) *A History of Conservation.* Oxford, Butterworth-Heinemann.

Larkham, P.J. (1996) *Conservation and the City.* London, Routledge.

Pickard, R.D. (1996) *Conservation in the Built Environment.* Singapore, Longman.

The Aga Khan Trust for Culture (1990) *Architectural and Urban Conservation in the Islamic World.* Geneva, Switzerland, The Aga Khan Trust for Culture.

Web-based sources

Council of Europe: www.coe.int

Europa Nostra: www.europanostra.org

International Centre for the Study of the Preservation and Restoration of Cultural Property (ICCROM): www.iccrom.org

International Council of Monuments and Sites (links to national committees and international scientific committees): www.icomos.org

Save Britain's Heritage (link to Save Europe's Heritage): www.savebritainsheritage.org

Society for the Protection of Ancient Buildings: www.spab.org.uk

The Aga Khan Trust for Culture: www.akdn.org/agency/aktc.html

The Getty Conservation Institute: www.getty.edu/conservation

The World Bank: www.worldbank.org

UNESCO (conventions, projects, World Heritage Sites, Intangible heritage): www.unesco.org/culture

World Monuments Fund: www.wmf.org

Chapter 3

Theoretical framework and conservation principles

There are many reasons for conservation; primarily buildings are preserved and conserved because they are useful and have value for their users, or because they testify to the identity of a national, ethnic or social group. In this respect, the built heritage can be seen as the physical manifestation of individual, collective and at times 'imagined' memory. Historic buildings not only provide scientific evidence of the past, but can also embody an emotional link with it, allowing an experience of space and place as it might have been experienced by others before us. Architectural heritage ranges from monuments that signify a long-distant victory for a nation to the vernacular style that is part of a familiar landscape. The splendour and the building techniques of ancient ruins may be viewed in admiration, while a townscape characterised by historic buildings brings to life a past way of living and can offer a sense of 'stepping back in time'. On a more familiar level, objects and places are links to an individual's past, a piece of furniture recalled from a grandparents' house, or a corner shop that is a familiar landmark associated with childhood memories.

The previous chapter highlighted some of the different approaches taken to conservation over the past two centuries. In each era, those leading the conservation movement firmly believed that they were doing the 'right' thing. Determining what is 'right' in conservation is not straightforward, either technically or philosophically, as both the knowledge base and values are constantly changing. Conservation is laden with conflicts and contradictions and there is no single, universally agreed method or methodology. Nor indeed should there be. Often it is the test of time that determines whether conservation has been successful or not. While success in technical terms can be judged against rates of decay and sustainability, the success of an approach or a philosophy can only be measured against the current values that society attributes to the cultural heritage.

This chapter discusses some of the core principles that underlie and inform conservation theory today. The first section, by way of background, identifies the values that are associated with the cultural heritage and defines the most commonly used terminology. The second section concerns the philosophy of conservation and outlines some fundamental principles and guidelines. These principles are re-examined in the second part of the book on practice in the context of altering buildings, structural stabilisation, approaches to the conservation of various materials and finally conservation and design in historic townscapes.

VALUES AND DEFINITIONS

Conservation is the process of understanding, safeguarding and, as necessary, maintaining, repairing, restoring and adapting historic property to preserve its cultural significance. Conservation is the sustainable management of change; it is not simply an architectural deliberation, but an economic and social issue. The concern of conservation is the past, present and future of a building and involves making balanced judgements in respect of:

- evidence (history)
- the present day needs and resources available
- future sustainability

A values-based approach

The significance of a building or place of historic, architectural and cultural importance is its most defining value, the loss of which will devalue its cultural significance. Yet, in a values-based approach to conservation, a much wider range of values also needs to be recognised, not all of which will relate to the physical fabric. Values are the qualities and characteristics that different users and different societies place on the cultural heritage at different times. In times of conflict, for example, cultural heritage often becomes a unifying symbol of identity; in other instances, cultural heritage can be deliberately exploited for political purposes. Cultural significance is made up and supported by a wide range of values, some of which may be in conflict with one another. Values most commonly associated with the cultural heritage are historic, architectural, aesthetic, rarity or archaeological values. Other values are less tangible and relate to the emotional, symbolic and spiritual meanings of a place. A values-based approach to conservation involves the recognition of the diverse range of values and responding to their needs through appropriate intervention and management. The role of conservation is to preserve and where appropriate enhance values.

Some buildings are built as monuments and continue to be valued in that way, others lose their intended value and significance in the passing of time, while some gain value for other reasons as society attributes new values to them and as they come to symbolise something else. The buildings and places valued as cultural heritage today are a reflection of current societal values, and may not be those that will be held by future generations. Furthermore, the values attributed to places of cultural, historic and architectural significance

Figure 3.1 The Parthenon in Rome has been a tourist trap for centuries, experienced by each generation with a sense of awe, yet most probably with different interpretations. The dome has continued to be a source of inspiration for architects to the present day.

may be held by people who have never seen or experienced them, and possibly never will. The knowledge value of an ancient site, for example, has an impact on learning across the world and not solely to the archaeological community or those visiting the site (Figure 3.1).

The values-based approach to conservation is an analytical method in which value judgements have to be as objective as possible. It is thus essential to both involve those representative of the different interests in the building or place, as well as the input from a multi-disciplinary team including anthropologists, social scientists and economists, all of whom will contribute different techniques and approaches to assessing values. Values need to be balanced against each other and prioritised when making conservation decisions. It is unlikely that all values can be treated with equal importance and accommodated within a project. In fact, some values will be in opposition to each other, requiring informed and balanced decision-making within the team.

Some of the values associated with cultural heritage are explained below. This list, presented in alphabetical order, is by no means exhaustive, but is intended as an overview and starting point when embarking on a conservation project. These values relate to all forms of architectural heritage including monuments, major public buildings, historic structures, vernacular buildings and historic urban areas. Not all values will be relevant to all places and their relative importance will alter from situation to situation.

Age and rarity value

Since the passage of time inevitably sees the loss of historic structures, the older a structure is, the more value is likely to be attached to it. What may be of value from one period may not be seen to have the same value from another. For example, a medieval workers cottage will be of value because there are few surviving examples, whereas nineteenth century workers cottages that are still very much in use and abundant in numbers are less likely to be afforded the same value and level of protection. Unlike buildings of more recent periods for which information is available in various forms (e.g. drawings, photographs, written accounts), no documented evidence exists on a prehistoric site. Therefore, the evidence transported in the physical remains is also arguably of greater value for these ancient structures for which there is no documentary evidence.

Rarity value can also relate to the occurrence of a building type or technique in an area where it is not commonly found. Some of the arguments against protecting and preserving twentieth century heritage have been that they have not as yet gained rarity value and that it is therefore not possible to identifying the exemplary buildings amongst hundreds of similar ones.

Architectural value

The exemplary qualities of design and proportion and the contribution that the architecture of a building has made to the quality of the everyday experience is its architectural value. In addition, the contribution the building makes to the architectural style or period, being the definitive work of a well-regarded architect or the use of pioneering building techniques, will also form part of the architectural value.

Artistic value

An artistic value may also be attributed to a historic building, linked to the quality of the craftsmanship or directly to artwork that is integral to the building, such as painted murals.

Associative value

The association that a building or place has with an event or personality in history is its associative value. The most obvious example of this is historic battlefields, where there is little if any in the way of physical evidence of the battle, yet the significance of the location cannot be denied. The loss of life suffered will add emotional and spiritual value, and the place of the battle in a nation's history will give it symbolic value, although this will most likely be interpreted differently by the two sides involved at the time.

Cultural value

Buildings provide information on various aspects of a past period, from lifestyle to the use of materials, crafts and techniques used in their construction. They may continue to play a role in current cultural traditions. In Indonesia, for example, many of the motifs used to decorate buildings from prehistoric times, continue to appear in batik and textile designs that are being produced today.

Economic value

The most highly regarded economic benefit of cultural heritage is tourism. By the end of the twentieth century, tourism had become a principal reason for the conservation of the cultural heritage the world over. However, alongside such direct economic benefits there are also less obvious economic values. There is considerable evidence of the built heritage contributing to the character and desirability of an area and the resulting increase in property values. The economic value of cultural heritage is discussed in more detail in Chapter 8.

Educational value

Historic sites and buildings have value in what can be learnt from them, with topics including a period of history, a past way of life, social relations or construction techniques. Educational value relates to a broad spectrum of learners from young children, through to life-long learning for all age groups. There is also an educational value attached to the conservation process itself. For example, the reason for building a replica building may be for its use as an educational tool.

Emotional value

People who use or visit buildings may feel an emotional attachment to them or may be moved by the building as a sense of wonder and respect at the artistic achievements in design and craftsmanship. This could range from an

Figure 3.2   History is quite literally embedded in the form of a Napoleonic cannonball in this wall in Bratislava in Slovakia.

emotional attachment to a local church and the memories it beholds to a feeling of amazement at the sight of a great building such as the Taj Mahal in India.

Historic value

A building or place is not only physical evidence of the past, but may also have played a role in history, or is linked to certain events or period in history. The history embodied within the building fabric is sometimes the only evidence to events and life in the past (Figure 3.2).

Figure 3.3 A historic town embodies values associated with quality of life, historic value, townscape and group value, local distinctiveness and the economic value for tourism.

Landscape value

The built heritage is an integral part of the landscape shaped by man. The appreciation and understanding of historic buildings has to include their context and setting. In some instances, buildings and landscapes constitute an artistic whole where the design of the monument, building or townscape and the landscape has been complementary. A good example of this is the English landscape garden of the eighteenth century, where ruins were incorporated into the designed landscape setting to become part of that aesthetic whole. Garden cities and suburbs of the nineteenth and early twentieth century were specifically designed to integrate domestic living with the landscape.

Local distinctiveness

Some of the value of a cultural heritage asset might be the contribution that it makes to the local distinctiveness of a place, providing a unique quality that makes it different from anywhere else. This could be the use of locally available materials or certain building techniques developed in the region. Historic towns are often valued for their distinctive characteristics in the face of repetitive and similar international styles of architecture (see Figure 3.3).

Figure 3.4   Wenceslas Square in Prague in the Czech Republic, acquired a new public meaning when Jan Palach, a young University student, set himself alight in the square in protest of the Russian occupation in 1969. (Photograph by Geoffrey Randell.)

Political value

Conservation cannot be separated from politics. The favouring of certain periods over others is often a political decision taken for a number of reasons. In Morocco, the authorities have little interest in preserving the art deco style buildings of Casablanca because they are associated with a period of French rule, and are not seen as part of the national identity, which identifies with architecture and monuments in the distinctively Islamic style. At the same time, it falls to the French Government to pay for the conservation of colonial period buildings in Laos. In other cases, politicians see the value of cultural heritage purely in terms of tourism revenues, which can alter the balance of values and distribution of funds for conservation.

Public value

Public spaces in particular will acquire public value, especially if they have been the scene of rallies, demonstrations or even revolutions, in which case they may also be regarded as being of political and historic value (Figure 3.4). Other places or buildings gain value in the public mind when they are threatened with demolition. Where the public has become organised and puts up a fight to save a building or place of historic importance, there will be a new public value attached to it. The power of public opinion on which buildings are safeguarded should not be underestimated.

Religious and spiritual values

For worshipers, churches, synagogues, mosques, temples and other places of worship embody a spiritual meaning and value. Not only places of worship,

Figure 3.5 A small square in Sevilla in Spain has social and public values for locals who regularly use it as a place of congregation, social exchange and play.

but also pilgrimage routes, nature in the form of mountains, rivers or other natural features are considered to have spiritual and religious value by different communities (see also Figure 2.6 of Uluru in Australia). There will also be spiritual value embodied in places that were once places of worship, but are no longer used for this purpose, such as where a pagan temple or redundant church has been converted to a new use.

Scientific, research and knowledge value

Whether it is the building techniques employed or the materials used, historic buildings have scientific values in terms of the information they contain on building practices of the period, which in turn inform conservation projects. They will contain valuable technical information of materials, where they were sourced, what type of tools were used in their construction and what caused their decay. Buildings may also contain evidence from past conservation interventions. The scientific, research and knowledge value is linked to educational value.

Social value

The meaning of a historic place to a local community, often as part of an ongoing social interchange, constitutes its social value. A local community may take pleasure out of using a local park or congregating in a local square, irrespective of its historic or architectural value (Figure 3.5). A central square may also be valued for its association with events and festivals.

Symbolic value

Erected to commemorate events in history, monuments will have intended and symbolic memorial value. However, that memorial value may change over time either through a change of political regime or simply by the sufficient removal of time from the event. Triumphal arches are now more likely to be seen as urban landmarks than symbols of past victories. The symbolic value of the Berlin Wall has been in constant evolution from the time of its conception, through to its demolition and reinterpretation in Berlin. It represents different symbolic values to different groups within the local population and to visitors, with each audience relating to it in their own way based on their own cultural values and understanding of past events. The symbolic value continues to shift as a new generation no longer feels connected to the Wall as part of their own living memory.

Technical value

The technological systems used in the construction of a building and its contribution to advancing building technologies at the time constitutes technical value. This may be the distance an arch spans or the use and development of a material that is new for its time. Technical value may also relate to the environmental systems incorporated into the design. Traditional techniques used to capture cool breezes in hot climates, such as the wind towers in the Gulf region, are being used to inform ecological designs practices today.

Townscape value

In many instances, it is not the individual attributes of a building, but its contribution to a group of buildings, street or townscape that is of value. As discussed above, buildings cannot be treated in isolation from their surroundings or settings. It is often the case that the group value is greater than the value of the individual components.

Terms and definitions

Conservation terminology can vary with language and according to the interpretation of different cultural communities. Even within the English-speaking world, some terms have different meanings in different regions. The following definitions are those commonly accepted in the UK and also aim to differentiate between various approaches to conservation. In the context of this book 'conservation' is used as an overarching term to include the intervention and management necessary to safeguard the cultural significance of historic buildings and their immediate environment.

Adaptive reuse/adaptation

Most buildings will change their use through their lifetime and this will invariably necessitate changes to the internal layout and fabric of the building. Making changes to a building to accommodate a new use is often a means of enabling the continued usefulness of a historic building. However, the appropriateness of the new use to the building fabric and its integrity does need to be considered.

Conservation

The Burra Charter (1999) defines conservation as: *'all the processes of looking after a place so as to retain its cultural significance. It includes maintenance and may according to circumstance include preservation, restoration, reconstruction and adaptation and will commonly be a combination of more than one of these'*. Conservation also includes cultural resource management and the management of change.

Consolidation

Physical interventions undertaken to stop further decay or structural instability. Some materials such as earth or mud require regular consolidation by their very nature as materials (after a severe rain storm for example). Capping of exposed wall tops is another form of consolidation, particularly in the case of ruins that are exposed to the elements.

Preservation

To maintain a building in its existing form and condition and undertaking maintenance work as necessary. Preservation is often used in American English in a similar way that the word conservation is used in British and Australian usage.

Prevention/preventative measures

To alter conditions to reduce or slow the process of decay. Preventative measures may include management of the internal environment, good housekeeping such as ensuring gutters and down pipes are not blocked or monitoring tree growth and vegetation around a historic property. No intervention to conserve or restore is involved.

Protection

Putting in place legal, physical or other tangible measures to safeguard cultural property from damage. Placing a historic building or site on a statutory list with legal restraints is a form of legal protection. Proactive measures in protection might include fencing around sites or the appointment of guards (Figure 3.6).

Reconstitution

Re-building a collapsed building or parts of it piece by piece. Reconstitution must be based on firm evidence and not conjecture. At archaeological sites in particular, it is likely that buildings would have changed, been added to and various components moved around over time, and hence there may not always be an obvious 'period' that can be reconstituted (Figure 3.7).

Figure 3.6   Placing a fence around a monument may safeguard it from vandalism, but without any intervention its condition will continue to deteriorate.

Reconstruction

Re-creation by building a replica of a building on its original site. Reconstruction is often undertaken to replace buildings or parts of buildings following fire, war damage, earthquake or other disasters. Although reconstruction must be based on sound evidence, it remains a re-creation and inevitably a reinterpretation of the past. Both authenticity and the patina and evidence of age are lost (see case study box on Uppark later on in this chapter). Reconstruction may be justifiable if a building is an integral part of a streetscape, square or complex and where its absence would detract from the integrity of the whole.

Moving buildings to a new site is also a form of reconstruction, often seen as a justifiable means of preservation if the building is threatened. In such cases, the context is lost as is other cultural evidence that may have been associated with the building. Recent cases of cultural sites being threatened by the building of major dams has involved emergency excavations followed by the relocation of some of the most outstanding monuments. Although the monuments are saved, they are exhibited out of context, the material belonging to the site is fragmented and any further archaeological evidence relating to the site is no longer accessible. It has also been the case that some building materials have reacted unfavourably to the new

Figure 3.7 This small altar has been reconstituted to where it was found in the precinct of the Roman Temple of Trajan at Pergamon. The altar, however, is known to be of an earlier date and later moved to the Temple precinct during the Roman period, to which the reconstitution relates.

environmental conditions, as has been the case with the Abu Simbel Temple in Egypt, which was moved to a new location when the Aswan Dam was constructed.

Open air museums are often seen as safe places to protect and display buildings that would otherwise be demolished (Figure 3.8). As such they have scientific and educational value, but the loss of context and setting needs to be considered in the interpretation. While timber frame buildings can be dismantled, transported and rebuilt with relatively ease in these situations, the same is not the case for buildings built out of more vulnerable materials like earth. This may affect the type and authenticity of buildings that are displayed.

Replicas

A replica is a facsimile copy of an existing building or part of it, often created for display purposes. There are cases where fragile and vulnerable stone carvings have been removed to safer museum conditions and replicas placed on the building. Such decisions need to be based on value judgements, where the artistic value of the carving is evaluated against potential threats of severe weathering, theft or vandalism. Replicas allow the integrity of the building to be maintained while preserving significant components and the artistic evidence that they contain and making them available for research.

Figure 3.8 An open air museum in Tokyo brings together vernacular house types from around Japan. The plan forms, however, have been altered in places to fit the curatorial need rather than maintain the building 'as found'.

Restoration

Restoration is returning a building or parts of it to a form in which it appeared at some point in the past. In the English language, restoration is often used synonymously with reconstruction. Although the word has become associated with major interventions of rebuilding to unify a building, small interventions such as replacing a missing detail is also restoration. The process where the objective of cleaning is to return a building back to its near original appearance is a form of restoration. Where restoration is necessary, it is paramount that all interventions are based on verified evidence. It is also important to provide sufficient differentiation between the old and new to avoid any misinterpretation in the future.

Some references to restoration in the Venice Charter of 1964 are more likely to be used as 'conservation' in current day usage and as has appeared in the more recent English language Burra Charter.

PRINCIPLES, PHILOSOPHY AND GUIDANCE

Two key international charters, the 1964 Venice Charter and the Australian Burra Charter revised version of 1999, embody much of the major guidance on conservation today. Through the ICOMOS International Scientific Committees, a series of more specific subject-based charters have emerged, and these are discussed and listed in Chapter 2 (see Table 2.1). Each of these charters builds on the general principles introduced by the Venice Charter. Another doctrine that has come to influence conservation practice is the UNESCO World Heritage Convention, the Operational Guidelines of which stipulate how significance is determined and how the cultural and natural heritage is managed. The convention is also supported by the 1995 Nara Document on Authenticity.

Each conservation project and problem will be different and must be assessed on its own merits against the identified values, demonstrating a sound and ethical approach that is guided by internationally accepted principles. The purpose of this section is to explain the basic principles that constitute conservation philosophy today. The first part concerns ethics in conservation and discusses the concepts of authenticity and integrity in conservation, whilst the second part focuses on the fundamental principles.

Ethics in conservation

Alongside, the basic principles are the underlying role of ethics in conservation. Conservation philosophy today advocates a values-based approach described above and is based on integrity and authenticity.

Integrity

Conservation must be undertaken with integrity, using materials appropriate for the purpose in a 'fitting manner'. A historic building is a relic from the past and holds details and information about the past; this is its historic integrity. Returning a building through restoration or reconstruction to how it is assumed to have looked in the past, for purposes of presentation or indeed authenticity, defies integrity.

Integrity includes:

- physical integrity (of the building materials and their relation to one another)
- structural integrity
- design integrity
- aesthetic integrity
- integrity of the building within its setting and context
- professional integrity of the conservation team

Authenticity

The Oxford English Dictionary defines authentic as 'genuine, of undisputed origin'. Other sources refer to authenticity as truth. There are many facets of truth or authenticity where a building conservation project is concerned, from the use of authentic materials to maintaining the truth in an architect's original design. Authenticity does not, however, mean original in the sense of returning a building to its original form. Conservation in many instances depends on interpretations of which there may be several, in which case there is not necessarily one truth.

In conservation, authenticity relates to:

- design or form
- material
- techniques, traditions and processes
- place, context and setting
- function and use

The World Heritage Convention, through its operational guidelines, requires sites nominated to the World Heritage List to be 'authentic' and refers to the Nara Document on Authenticity for guidance in this respect. The Nara Document states that: '*All judgements about values attributed to heritage as well as the credibility of related information sources may differ from culture to culture, and even within the same culture. It is thus not possible to base judgements of value and authenticity on fixed criteria. On the contrary, the respect due to all cultures requires that cultural heritage must be considered and judged within the cultural contexts to which it belongs*' (Nara Document on Authenticity, 1995).

Replicas are fakes and lack authenticity in many respects. Authentic material is the only concrete evidence of history that can be carried into the future. As much as buildings can be recorded and these records retained in carefully maintained archives, the informative value of original material cannot be fully replaced. In investigative terms, new technologies are constantly becoming available that will allow original material to continuously reveal new information. However, while the genuine article may be of most value to Western interpretation, in other cultures the sense of place may be much greater than the material value of the built form. In such cases authenticity resides in place, design and the spirituality of place more than in material relics.

In some cases, material and aesthetic authenticity may contradict one another. Viollet-le-Duc's restorations to a Gothic ideal, discussed in Chapter 2, are not authentic in that he was creating an imagined aesthetic unity and disregarding the buildings' history and development. In the case of restoring a modern movement building, the retention of the aesthetic of a monolith surface by fully replastering instead of undertaking localised patch repairs, respects the authenticity of the original design over the authenticity of the material (Figure 3.9).

Principles of conservation

While an overall project philosophy and approach will be determined for each project, there will be thousands of small decisions on individual interventions and material decisions that need to be made as the project progresses

Figure 3.9   Mies van der Rohe's Barcelona pavilion, whilst being a source of inspiration to the hundreds of architecture students who visit and photograph it each year, lacks authenticity in many respects. It is a replica of the original, adapted for the purpose of being a permanent building, and in this sense it is not authentic Mies. Furthermore, it is now being displayed out of its original context.

on site. When taking these decisions, it is incumbent on the professional to consider the overall project philosophy and conservation principles in general as well as considerations of integrity and authenticity. This section outlines the basic principles of conservation as they are interpreted at the current time. The principles apply to the overall approach as well as individual interventions. They have been grouped under the three headings of understanding, implementation and evaluation as explained in Table 3.1.

Table 3.1   The basic principles of building conservation.

Understanding	Working with the evidence
	Understanding layers
	Setting and context
Implementation	Appropriate uses
	Material repairs
	Tradition and technology
	Legibility
	Patina of time
Evaluation	New problems may require new approaches
	Sustainability
	Interpretation

Working with the evidence

'The process of restoration is a highly specialised operation. Its aim is to preserve and reveal the aesthetic and historic value of the monument and is based on respect for original material and authentic documents. It must stop at the point where conjecture begins, and in this case moreover any extra work which is indispensable must be distinct from the architectural composition and must bear a contemporary stamp. The restoration in any case must be preceded and followed by an archaeological and historical study of the monument' (Article 9, The Venice Charter).

It is often very tempting to rebuild on the basis of supposed but not proven evidence. Conservation, however, must be based on verified evidence and no attempt should be made to replace fabric where there is no firm evidence backing such conjecture. Nor should historic evidence be falsified as a result of interventions. The smaller the intervention, the more appropriate it is likely to be. For example, to achieve integrity and functionality a missing part of a doorway may be completed based on the evidence of the remaining in situ section. Furthermore, most buildings will have evolved and changed throughout their life and returning details to a certain chosen period can be problematic and contradictory. Replacements and restoration should not be at the expense of later but significant additions to the building. Where details are not known interventions should either be in a contemporary style or 'in the spirit of' the historic detailing, but clearly distinguished as a later addition.

Case study: Uppark – rebuilding after the fire

Uppark, and Windsor Castle are two notable historic properties in England that suffered significant fire damage in 1989 and 1992, respectively. The post-fire restoration in each case was keenly debated by leading conservation professionals and academics.

At Uppark, the unprecedented step was taken to rebuild the property back to 'the day before the fire', a decision that was partly influenced by the conditions of the insurance payment. The reason cited was that this was the best way of preserving the integrity of the landscape and the contents of the house, most of which were rescued from the fire. The house, of which only the shell remained, was rebuilt and interior finishes refitted to appear in the condition they were prior to the fire. In places, the wallpaper was specially made to appear faded in places where it had been exposed to sunlight. While the work of the conservators is commendable and the integrity of the property has been maintained, questions need to be asked about how the materials are going to age from the time they were put back to, and whether the patina they are going to acquire will be the same as what might have been had there not been a fire.

Building materials will also contain evidence of the past for which techniques of analysis may not have been developed yet. Modern techniques including microscopic studies provide opportunities for in-depth analysis of material components and make-up and it is important not to destroy evidence that may be an informant in the future.

Understanding layers

'The valid contributions of all periods of the building of a monument must be respected, since unity of style is not the aim of restoration. When a building includes the superimposed work of different periods, the revealing of the underlying state can only be justified in exceptional circumstances and when what is removed is of little interest . . . ' (Article 11, The Venice Charter).

Over time, there are likely to have been numerous changes to the fabric and layout of a building, including conservation and repairs (see case study box on Gloucester Blackfriars in Chapter 5). In places, substantial additions may have been made and/or removed. Through various conservation interventions over the years a plethora of methods and approaches may also be evident, some of which would not be acceptable to present-day conservation and interpretation principles and practice. Many of these interventions will have become a layer of the historic fabric and are informative of their own time, and should be recognised as such. Nonetheless, there are times when the removal of historic fabric is justified, particularly where later additions are causing damage to older original fabric or detract from the cultural significance and integrity of a building or place. This might be the removal of a low quality utilitarian extension or a later render that is trapping moisture in the building fabric.

Development proposals will need to recognise the totality of the buildings on the site, including the evidence and fabric existing above and below ground, the various development phases and the significance of the links that exist between each individual section. Conservation is simply another layer in the ongoing history of a building and any new intervention and ongoing maintenance, repair and servicing works should be recorded.

The archaeology of a site should be recognised as belonging to the wider context of archaeology in the locality. Invisible and below ground layers of a site should not be ignored and may need to be protected, especially when planning underground works such as drainage and other utilities.

Setting and context

'A monument is inseparable from the history to which it bears witness and from the setting in which it occurs. The moving of all or part of a monument cannot be allowed except where the safeguarding of that monument demands it or where it is justified by national or international interest of paramount importance' (Article 7, The Venice Charter).

Monuments have often been designed as an integral whole with their landscape or to make a specific impact on a setting, such as a cathedral rising from within the close knit low-rise and dense urban fabric of a medieval city. Not only monuments but other forms of architectural heritage are also integrally linked to their setting and surrounding landscape, whether it is the relationship of a house to the context of a historic city or street in which it is located, or the link between an abandoned mill and a river or waterway it once depended on for its operation and the transport of products. The setting and surrounding of historic buildings may have undergone numerous changes over time and this change is likely to continue. However, the guardians of a historic property may have limited influence over future developments surrounding the site. Approaches and decisions concerning conservation should not isolate a building from its setting.

Appropriate uses

'The conservation of monuments is always facilitated by making use of them for some socially useful purpose. Such use is therefore desirable but it must not change the lay-out or decoration of the building. It is within these limits only that modifications demanded by a change of function should be envisaged and may be permitted' (Article 5, The Venice Charter).

With the exception of major monuments of outstanding historic and symbolic significance, most buildings become redundant when they are no longer suited for the use they were built for, or when the users needs have significantly altered to the extent that the building is no longer fit for purpose. If buildings cannot be adapted and changed to accommodate new uses, then they will be abandoned and interest in their conservation is less likely to be forthcoming. Even though a new use is a means of enabling continuity and reuse, the proposed use must be appropriate to the fabric and layout of a historic building, and most importantly must not detract from its cultural significance. The use of the site in the present day or in the future must continue to enhance the archaeological, architectural and historic value of the site. An industrial or polluting use, for example, is unlikely to be desirable, as are uses that will impose heavy strains on the building fabric or threaten its integrity.

Material repairs

'Conservation is based on a respect for the existing fabric and should involve the least possible physical intervention. It should not distort from the evidence provided by the fabric' (Article 2, The Burra Charter).

The old adage that 'it is better to maintain than to repair, better to repair than to restore and better to restore than to rebuild', a dictum supported by William Morris in the SPAB manifesto, is a useful starting point in assessing approaches to conservation and a reminder that 'less is often more' in conservation.

Apart from monuments built to last for posterity, much of our utilitarian and domestic building stock was built for shorter life spans. In some instances, conservation has been able to provide them with a much longer life than their original builders would have intended. In Turkey, parts of the Middle East and the Balkans, timber frame houses built in the eighteenth and nineteenth centuries were only expected to last one to two generations, at which point they would be renewed to fit the needs of the evolved extended family that occupied them. For this reason, they were often constructed in low quality materials. Consequently, their conservation needs can become substantial.

In the twentieth century, rapid urbanisation and economic pressures on city centres the world over has seen continuous replacement as a response to ever-increasing property values. In places, centrally located commercial buildings are being replaced as often as every 20 years with newer and taller versions. Even the environmentally conscious buildings that are being conceived now are rarely constructed with the inbuilt flexibility that will allow them to be easily adapted and updated for future uses. Thus, even the design intention of buildings that are listed from the more recent past might have been for a relatively short life span that could create potential problems for their repair and maintenance.

The level of intervention will depend on the value in the buildings, both as an entity and in its material. For example, in the case of a seventh-century Saxon church, each stone is important for the evidence of masons marks, use of tools and the like, whereas a mid-twentieth century cinema building is more a symbol of its time, but has less value in terms of materials when it comes to proposing interventions.

The aim of repair is to slow down the process of decay. However, present day health and safety regulations require the use of more scaffolding for access than previously. In view of the high cost of temporary structures and site set up in proportion to the cost of repairs, there will increasingly be a tendency to carry out a substantial amount of work, including replacements in anticipation of future defects and failure, in order to avoid access costs over the next few years. In most projects, a careful balance will need to be sought to ensure the building remains secure for sometime, but its historic integrity is not threatened in doing so (Figure 3.10).

Tradition and technology

'*Conservation should make use of all disciplines which can contribute to the study and safeguarding of a place. Techniques employed should be traditional but in some circumstances they may be modern ones for which a firm scientific basis exists and which have been supported by a body of experience*' (Article 4, The Burra Charter).

Wherever possible, repairs should be carried out following original building techniques, except where these are found to be the cause of decay or failure. Using traditional methods should be the first option especially where it is apparent that the continuation of traditional methods is also maintaining the tradition of the building and its link to the local community. Where knowledge and skills continue to be passed down through generations of craftsmen, extensive research or testing of materials can also be avoided.

Using the same materials as surrounding original material will ensure that the building element (e.g. wall, roof truss) will continue to behave and move in the same way. Sometimes, however, new solutions or materials are the appropriate answer to old problems. New materials should only be used for repairs if they have proven over time to be effective and not damaging to

Figure 3.10 At Canterbury Cathedral, England, a previous repair with a concrete wall plate and the roof timbers embedded in it poses a problem for new roof repairs.

the historic fabric. The commonly agreed principle that all conservation work will be reversible is not always the case and in some instances not possible; much of the work carried out using modern materials, adhesives and resins is not reversible.

Most buildings will have been repaired over time, using a number of different methods and approaches. These contribute to the life story of the building and should not be discarded or dismissed without serious consideration. It is also important to maintain consistency in attitude and implementation of repair, especially where this is needed to sustain the performance of certain materials. Therefore, some conservation decisions may have to be taken in respect of previous work. Equally, decisions taken and methods chosen should not prejudice or hinder future works. However, where later conservation works are proving to be detrimental to the preservation of the earlier fabric and they have been undertaken as 'reversible' interventions, then their removal may become necessary, so long as the removal process is no more damaging to the fabric than the conservation work itself and that a more sympathetic option is available. For instance, even where a hard mortar pointing is compounding the erosion of the surrounding masonry, its removal may be even more damaging. A cement-based render is almost impossible to remove without causing undue damage to the masonry base.

Figure 3.11 At the Temple of Teotihuacán in Mexico small stones in the mortar joints has been used to differentiate between original structures and new work, but this may only be legible to the discerning professional eye.

Legibility

'*Replacements of missing parts must integrate harmoniously with the whole, but at the same time must be distinguishable from the original so that restoration does not falsify the artistic or historic evidence*' (Article 12, The Venice Charter).

While the first concern of conservation is the safety of the existing structure, this will invariably be followed by aesthetic considerations. One of the biggest challenges for conservation professionals and conservators working on site is the visual appearance of a repair in relation to the whole. Choices have to be made among differentiating a repair from the original, maintaining an overall aesthetic and the different levels of weathering apparent on old and new materials (Figure 3.11). Where a building is constructed from various materials, both in its original and subsequent building phases, the introduction of yet another material to highlight new additions may become more misleading than it is informative. On an urban scale, conservation will have a visual impact in terms of townscape character.

'Honest repairs', as advocated by the SPAB, have been much debated, especially regarding issues of integrity. Years of honest and apparent repairs may start to detract from the character and integrity of a historic building. Inevitably, a new piece of work, whether a brick replacement or mortar pointing, will appear new on completion and contrast with surrounding work. Although this may not be visually appealing initially, natural weathering will make it much less apparent as time goes by. The use of techniques that make the new materials appear to be old move away from the honest portrayal of the intervention.

In the West, one repair approach has been not to reproduce extensive decorative mouldings but to present new sections with a simplified or stylised profile. On the other hand, in cultures such as in India, where craft traditions are still alive and thriving, it is often considered acceptable to continue the tradition of full detail carving in repair works and recognise that conservation includes the conservation of the craft tradition as well. In some European Cathedrals, in time-honoured tradition, stone masons have been given a free hand in carving new details, such as a gargoyle, where an original has been missing.

Figure 3.12 Historic monuments, like the Chini Mahal in Daulatabad, India, are likely to have undergone several stages of development and change, including repairs. Part of their present day integrity lies in the evidence and marks of these different periods as they appear in the building's fabric.

Patina of time

Historic buildings and the materials they are constructed with will weather and decay over time. The evidence of age, often referred to as patina, is also one of the values of a historic building or townscape. Buildings acquire other scars over time as well, from small scratches and chipped arises to shrapnel damage inflicted during times of conflict. The objective of conservation, however, is not about returning buildings back to a 'good as new' state or fighting a natural aging process (Figure 3.12). Cleaning building surfaces, whilst at times a necessary step to reduce the rate of decay, will if undertaken too rigorously not only compound decay but also remove valuable evidence such as masons marks, tooling and finishes from surfaces. Various methods for cleaning historic buildings are discussed in more detail in Chapter 7.

New problems may require new approaches

All the above principles continue to be challenged, as conservation professionals are faced with finding appropriate solutions for new types of conservation problems. This includes the consideration of intangible values (see Chapter 2), and the conservation of some more recent period architecture, notably the industrial heritage and some modern movement buildings where the authenticity of design concept can at times override material considerations, or where materials have not performed adequately for the purpose.

Not only are many late twentieth century buildings still seen as part of the current building stock in terms of use, they have not as yet established their position in nostalgic memory either. Owners and users of twentieth century buildings do not expect them to be protected nor the opportunity for alterations to be restricted. This, however, is only one of the challenges facing modern movement buildings. Unlike traditional buildings that were built from a relatively limited palette of materials for which conservation solutions and techniques have been developed and established, from the middle of the twentieth century the number of materials available to designers has grown exponentially. Not all have been successful and some have either been improved or abandoned. Where such materials have been used on buildings that are now listed, then conservation professionals need to consider appropriate solutions to their repair that maintain the intended character, but ensure adequate performance as well.

Another dilemma is faced with former industrial and working buildings, where the only means of repair to operable condition is to replace components. Increasingly, historic ships in dry dock conditions are listed and treated like historic buildings in terms of legislation and policy framework. Approaches to their conservation and maintenance may, however, need to vary from that of a historic building. A historic ship, for example, does not consist so much of a series of layers as its use has evolved, but the necessary and sometimes substantial renewals or additions to maintain it as a seaworthy vessel.

Sustainability

Conserving and reusing an existing building is in itself a more sustainable approach than complete renewal or replacement with a new building. Many traditional building practices were sustainable in that buildings were repaired frequently but with small interventions. Re-thatching was often the case of only replacing the topmost layer of a built up roof, using material that was grown locally in sufficient supply to allow for regular replacement. The choice of minimum intervention and regular maintenance of historic buildings is an ecological and environmentally sensitive approach to building conservation.

Conservation decisions also need to consider sustainability in terms of sourcing materials and this may conflict with good conservation practice. Considerations include the sustainability of forests in sourcing timber (especially some tropical hardwoods including mahogany), the availability of stone from quarries and the impacts of quarrying on the environment. Many timber structures were built at a time when timber was available in abundant quantities, creating problems for sourcing economically for conservation purposes. Avoidance of using harmful chemical substances for cleaning and conservation is also a more environmentally sensitive approach, although there are times when the use of chemicals is unavoidable. Nonetheless, there has been a tendency to seek 'miracle' solutions to prevent or halt decay, whereas allowing for the lifecycle of decay may be a more ecological or sustainable approach (see case study box in Chapter 2 on Cappadocia).

Interpretation and conservation

The conservation principles discussed in this chapter are well known to conservation professionals and practitioners, but the interpretation of their outcome by a lay audience may be very different. It is the role of conservation professionals to consider how conservation practice will influence the way that historic buildings and monuments are experienced, especially as most human interaction with the historic environment is incidental. Although developing interpretation policies for a site are not the remit of conservation, interpretation is an integral part of conservation and conservation decisions will inform interpretation.

Interpretation is the art of presenting the cultural significance of a building or place to its users, visitors and wider community. Where there is an identified desire to make the monument accessible to the general public, and where a building is specifically conserved for public consumption, then it is more likely to be interpreted for this purpose. A strategy for conservation will also relate to the management and interpretation of the site. Although the interpretation of historic buildings and sites might be seen as the domain of tourism and visitor attractions, the methodologies and approaches adopted for the conservation of these buildings and sites will play an important role in how they are interpreted and understood by visitors. The approach to conservation may need to be communicated alongside the meaning of the building.

The common practice of 'honest repairs', for example, dating back to the nineteenth century will not necessarily be obvious to visitors, probably at times giving the impression that the use of two different materials (e.g. stone and brick) was a common building method of the time, as indeed it is in places. Especially, when the repair itself is old and sufficiently weathered distinctions between new and old become even more obscure. Furthermore, each generation has had its own approach to repairs and their presentation, so gaining unity or common language on a monument that has been repaired at intervals over a century is difficult or most likely impossible (Figure 3.13).

In a museum or visitor attraction, there will be a desire to create the experience of living in another era, where restoration rather than identifying or differentiating intervention, attempts to blend it in. Reference is often made to providing an 'authentic experience' for the visitor, offering an insight into what the building would have looked, smelt and sounded like in its original use.

Careful consideration also needs to be given to areas where a balance is required between the needs of interpretation and the interest of the archaeological material. These choices should be regularly reviewed in relation to the

Figure 3.13   The forum in Rome, Italy, has been the object of conservation programmes since the 19th century with no discernable unity in approach.

condition of the ruin and its state of preservation. In the case of consolidation of exposed ruins, heavy intervention like capping or placing protective structures over ruins can detract from the understanding of the element and its overall link to the site. Today, digitisation, virtual reality, recreation of historic buildings and places as walk-through experiences are a useful tool for conservation professionals in understanding a building and various stages of its development and what may or may not have been possible in the original construction. Equally, these technologies give those visiting, including on a web page, the opportunity to see how a place may have looked like at a certain time in history, thus reducing the need to replicate, reproduce or restore simply for the benefit of the visiting public.

SUMMARY AND CONCLUSION

Although many of the principles discussed above have been established for over a century, they are also continuously evolving to reflect the values that current day society places on cultural heritage. The philosophy and principles, nonetheless, provide an essential framework for the implementation of conservation projects, whether they are small-scale interventions linked to maintenance or an outstanding new extension to a historic building. In summary:

- Conservation must be based on an understanding of the historic development of a building or place, its cultural significance and a wide variety of values attributed to it. Where there is doubt, it is advisable to undertake the minimum to stabilise until further information becomes available.
- Material repairs should follow well-recognised principles and be based on a professional understanding of the material qualities and the causes of decay.
- An ethical approach to conservation must be based on integrity and authenticity.
- The principles will remain a source and guide, but each situation will need to be judged on historic information and the current realities of the situation.
- Conservation concerns the past, the present and the future (sustainability).

Many of the basic principles of conservation outlined in this chapter also form the basis for conservation policy and legislation, which is discussed in the next chapter. In practice, it is not always possible to abide by all the principles all of the time and there will be occasions when decisions will need to be prioritised and compromises made. Chapter 5 provides an overview on how to approach a conservation project, while Chapters 6 and 7 provide a guide to structural repairs and approaches to the use of various materials in relation to these principles.

FURTHER READING AND SOURCES OF INFORMATION

Australia ICOMOS (1999) *The Burra Charter.*

Binney, M. and Hanna, M. (1978) *Preservation Pays: Tourism and the Economic Benefits of Conserving Historic Buildings.* London, Save Britain's Heritage.

Brown, A. (1998) *Windsor Castle Fire and Restoration.* Windsor, Cobblestone Communications.

Clark, K. (2001) *Informed Conservation.* London, English Heritage.

Earl, J. (1996) *Building Conservation Philosophy.* Reading, College of Estate Management.

Feilden, Sir B.M. (2003) *Conservation of Historic Buildings*, 3rd edn. Oxford, Architectural Press.

ICOMOS (1964) *The Venice Charter.*

Pickard, R.D. (1996) *Conservation in the Built Environment.* Singapore, Longman.

Rowell, C. and Robinson, J.M. (1996) *Uppark Restored.* London, National Trust.

Teutonico, J.M. and Palumbo, G. (eds) (2000) *Management Planning for Archaeological Sites.* Los Angeles, CA, J. Paul Getty Trust.

UNESCO (1994) *The Nara Document on Authenticity.*

UNESCO (2005) *Operational Guidelines for the Implementation of the World Heritage Convention.* Paris, World Heritage Centre. Available at www.unesco.org/culture

Web-based sources

Getty Conservation News: www.getty.edu/conservation/publications/newsletters/index.html

ICOMOS website for charters: www.icomos.org

Chapter 4
Legislation, Policy and Guidance

Cultural heritage is continuously threatened by development, rapid urban growth, globalisation, warfare, natural disasters and everyday neglect. Policy and legislation are the legal mechanisms for protecting the cultural heritage and controlling development. Legislation seeks to regulate and control what is permissible, and places limitations on change and development, whereas policy guidance provides advice on conservation issues. It is not possible in the confines of a single chapter to address the very different systems that are in place across the world. The intention here is to provide an overview of the type of systems that are most commonly seen and illustrate it with examples from Britain.

The previous chapters discussed the recent paradigm shift in heritage protection from sites of archaeological interest and monuments to encompass contextual issues, group value and areas and places of historic, cultural and architectural significance. Legislation regarding the protection and management of cultural heritage, however, is still in the process of adapting to these changes and does not always offer, for instance, spatial planning policies for areas of historic significance or indeed integrated approaches to implementation.

Despite the large number of signatories to international charters and conventions, both the definitions and the approaches to the protection of the cultural heritage continue to vary significantly across the world. While some of the variation is captured in legislation and policies, much is linked to the tools through which the legislation controlling the protection and conservation of historic buildings and places is applied, and how development is controlled. This control is often exerted through a planning system, in which most of the key players will interact with one another through the consent process.

This chapter is in two sections. The first identifies the various players responsible for decision making in the protection, conservation and management of the built heritage. The second section reviews the various levels of protection and control that apply to different categories or groups of cultural heritage.

Table 4.1 Levels of decision making for the protection of the built heritage.

Type of site	Level where decisions are made	Agency responsible
Archaeological sites and ancient monuments	National	Ministry
World Heritage Sites	National/international	Ministry UNESCO
Monuments and buildings of national importance	National	Ministry and/or its appointed agencies
Other listed/protected buildings	Local with national level input	Local authority with input from Ministry and/or its regional offices and appointed agencies
Conservation areas	Local	Local municipality
Areas of influence, setting of a historic building or place	Local/national	Local planning authority
Historic parks, gardens and areas of landscape significance	Depends on which legislation it is covered under	

DECISION MAKING IN CONSERVATION

Decisions concerning the cultural heritage or about factors that will impact on it are made at many levels and under differing circumstances, with different working relationships between the various decision-making bodies. The following section considers some of the key players who influence decisions concerning the protection, conservation and enhancement of the cultural heritage. Even the legal framework of decision making can be subject to conflicts. This conflict is most prominent between various government departments and between the priorities of national level protection priorities and local economic and community development agendas. Table 4.1 summarises the levels of decision making linked to the different types of cultural heritage and the agencies that are likely to be involved.

International level

Conservation law, policy and legislation varies considerably from State to State, as does the organisational structure that governs its application. Even within Europe, approaches to conservation and management of conservation vary considerably.

There is no international law governing conservation. By signing up to international conventions, State parties indicate their intention to follow the code of practice set out by the convention either voluntarily or through

national legislation. For example, signatories of the UNESCO Convention for the Protection of Cultural Property in the Event of Armed Conflict (the Hague Convention, 1954) undertake to respect the cultural heritage of all parties in their actions in conflict situations.

Cultural sites designated as World Heritage Sites, in accordance with the UNESCO World Heritage Convention, are not directly protected by UNESCO, but by the conservation, planning or environment policies of the country in which they are located. Unless a State has specific legislation concerning the protection and management of World Heritage Sites, then they are protected under existing legislation reflecting their status as a protected archaeological site, a listed building or an urban conservation area. Another role of the international community will be to provide technical and financial aid and at times put pressure on governments to encourage the protection of historic monuments thought to be under threat. Under the terms of the convention, the World Heritage Committee may also consider the removal of a site from the World Heritage List, if there is sufficient evidence that the significance and values for which it was inscribed are being eroded or lost. To date one site has been removed from the list.

Two European Conventions from the Council of Europe: the Convention for the Protection of the Architectural Heritage of Europe, the Granada Convention, of 1985 and the European Convention on the Protection of the Archaeological Heritage, the Malta Convention, of 1992, are significant as being Europe-wide conventions. The Granada Convention advocates integrated conservation practices and decision-making that was first launched in the Amsterdam Declaration of 1975. The definitions and categories introduced by the Granada Convention have in only a few cases been incorporated into the national legislation of its signatories. The convention also advocates integrated approaches to conservation and better cooperation between conservation, town planning and development agencies, a process which can be seen to varying degrees in member countries. Although there are as yet no specific Europe-wide policies on cultural heritage, several other legislative areas impact on cultural heritage and its conservation. Europe-wide standardisation of building codes will particularly concern the conservation of materials.

National level

The protection and conservation of cultural heritage is generally overseen at national level through a designated Ministry supported by legislation regulating statutory designation, protection and possibly development control. In different countries, the jurisdiction for antiquities may fall under a number of Ministries, including Culture, Environment, Education or Tourism. A Ministry of Environment will encompass the protection of natural sites as well as architectural sites in its remit. There are a growing number of countries where Culture and Tourism are brought together in a joint ministry (e.g. Turkey, Jordan, India), in recognition of the role that cultural heritage plays for tourism. In France, the Ministry of Culture is also responsible for architecture and is keen to encourage innovative design in the context of historic buildings. In some countries, cultural heritage belonging to different periods is regulated by different departments or even ministries. In other cases, a distinction is made between movable and immovable cultural property. Urban conservation and the protection of urban or rural historic areas are more likely to be managed through a Ministry responsible for planning, municipalities and rural affairs though the established town planning framework. Raising public awareness on the value of cultural heritage is generally undertaken at national level, and might also involve national media organisations.

Other Ministries will also have decision-making powers that impact on the protection of the cultural heritage. The Ministry of the Environment, for example, may be responsible for controlling areas that are significant to the setting of historic places or properties. The planning and management of major infrastructure projects from roads and dams to irrigation, energy provision including sources of renewable energy may impact on the cultural heritage. The Ministry of Finance will be responsible for the allocation of funds for the effective protection, management and conservation of the cultural heritage. In some countries, the law may exempt certain properties from the antiquities and conservation legislation, such as those belonging to the Ministry of Defence or sometimes religious buildings.

Depending on the size of a country, ministries will be represented at regional or local level through their appointed agencies or through their obligations under the planning system which is likely to be managed by local municipalities. In France, the Ministry of Culture has regional directorates, whereas in Germany the semi-autonomous regional governments of the Länder regulate cultural heritage protection and conservation. Statutory lists are often generated at national level and decisions concerning monuments and important historic buildings are more likely to be taken at national rather than regional or local level. There is often a debate on how much power should be exercised locally, with the response governed by the size of a country and national regulatory framework that is in place. There are varying models from exclusively national legislation, such as in Italy, to national legislation that is supported by regional legislation, such as in Spain, or legislation that is largely regional, such as in Germany.

In England, the Department of Culture Media and Sport is the government Ministry responsible for cultural heritage. The Historic Buildings and Monuments Commission for England, known as English Heritage, is an executive non-departmental public body that is appointed as the government's statutory advisor on the historic environment in England. A similar role is undertaken by Historic Scotland and Welsh Historic Monuments, Cadw, in Scotland and Wales, respectively.

Local level

At local level, conservation is often an integral part of the planning process where decisions are regulated and influenced by planning officers and elected members. The obligations of a local authority planning department will include reactive actions in the form of development control as well as proactive actions such as regeneration and environmental improvement initiatives. Both will impact on conservation. As at national level, the work of other municipality departments will impact on historic buildings and areas, including planning and development, transport, infrastructure, waste disposal, social services and public welfare.

Any form of development, including a change of use, will require consent from a local authority planning department. Often referred to as the planning process, this involves formal application by a building's owner for approval by the planning department. The role of the planning department is to review the application within the framework of its own urban plan, development objectives and historic area policies. Decision making at local level where the cultural heritage is concerned is often undertaken in consultation or in conjunction with national level heritage departments or their regional representatives. In larger municipalities or historic towns, one or several members of the planning team may be specifically dedicated to manage conservation issues. In England, many local authority planning departments have dedicated Conservation Officers whose role it is to specifically advise on applications regarding listed buildings and conservation areas, as well as work on projects concerning the enhancement of historic areas.

In most northern European countries, town planning is seen as the basic tool for heritage protection, even when it is not combined with specific legislation. This is because they have well established and efficient systems in place. In local town plans, land use policies should ideally be integrated with conservation policies. Planning is a broad discipline, in which conservation and conservation areas and the concern for historic buildings is but one aspect. A more effective approach to protecting, conserving and using historic buildings is not to see them as a separate 'add on' category, but as an integral part of the city plan, character and value of a place. To achieve this, integrated planning approaches and good communication practices between various departments and ongoing dialogue with private and non-governmental sector partners is essential. In Tunis, Tunisia, for instance, the Old Town Preservation Department has played a key role in planning, social welfare and capacity building as part of its efforts to conserve and regenerate the historic quarter.

Local municipalities also play a pro-active role in promoting historic areas and their conservation through the development of town plans and development policies. Potential new uses for a redundant building can be informed by the local authority in relation to plans and development frameworks being developed for an area or even the city as a whole. A longer-term vision, rather than a short-term solution, will be in the best interest of the building and the money that is being invested in its regeneration. Some authorities have also produced 'design guidelines' to inform conservation and new build projects in historic areas (see Chapter 8). The Penang City Council in Malaysia, for example, has produced Design Guidelines for conservation areas of Georgetown.

Non-governmental organisations and amenity societies

There will often be a number of non-governmental organisations and advisory bodies at national and local level that provide advice and support for conservation. Some will have been specifically established to undertake conservation projects, while others provide more research, support or advisory roles. The concern of these bodies may also be a certain type of cultural heritage or period in history. In most cases, amenity societies have embraced a lobbying or campaigning role, either on a single issue or for the general preservation and enhancement of historic buildings or area of historic significance.

Alongside the established societies of national standing, there will be numerous smaller societies and groups concerned with local issues. These play an important role in promoting and safeguarding local cultural identities and may sometimes function as advisory committees to municipality planning departments. In some places, it is a statutory obligation to consult some of these advisory bodies. In England, for example, amenity societies that are consulted during the planning process depending on the nature of the application include the Society for the Protection of Ancient Buildings (SPAB), the Council for British Archaeology, the Georgian Group, the Victorian Society and the Twentieth Century Society. Where parks and gardens are concerned the Garden History Society may also be consulted. The ICOMOS national committee, ICOMOS-UK is consulted on applications concerning World Heritage Sites.

In England, the Civic Trust, established in 1957, plays an important role in promoting the value of high quality environments as 'better places for people to live and work in'. Private and charitable trusts can also own and manage historic buildings. The National Trust, established in 1895, has led the way for many similar trusts to be established across the world, and today owns and manages over 200 historic properties in England, often setting standards for best practice. In more recent times, the Landmark Trust has played an important role in rescuing and conserving derelict properties and letting them as holiday accommodation. Similar organisations have appeared in other places such as Morocco, demonstrating the potential economic value of historic properties in tourism destinations in particular.

Civil society is increasingly playing a greater role in decision making or influencing decisions as well as providing support for activities and initiatives that cannot be supported by the State, from education and training through to providing grants for conservation and improvement works. Another important role played by amenity societies is in raising awareness and interest in the historic environment and its protection.

Alongside national and local organisations, there is also an international network of non-governmental organisations that are involved in conservation. Some act in an advisory capacity, others directly or indirectly provide financial or technical support, and some lobby. Those who play a role in financing conservation are noted in Chapter 5.

The private sector and property owners

The owner of a building is undoubtedly a key player in decisions that are taken in relation to the care, conservation and adaptation of their property. Ownership is linked to responsibility, and any statutory form of protection or listing cannot be separated from ownership. By placing a building under some form of legal protection, the State authority is invariably burdening an owner with new responsibilities to protect and maintain the property in a certain way, as well as placing constraints on change and development. In most cases, and certainly in poorer countries, the State is generally not in a position to provide the additional finances needed to fulfil these requirements. While in some places the official recognition of the historic value of the property is seen to increase its economic value, in other instances it is seen as an additional burden and even a hindrance to development.

Justifying the value of retaining and conserving a historic building can be problematic, especially when the land it occupies is considered to have higher development value. In prime city centre locations where land values are high, the bargaining power of a private developer can often influence the way decisions are taken, whether it concerns existing buildings, the redevelopment of vacant plots or the replacement of historic buildings. Over time this will change the character of a place if allowed to proceed unchecked. Decision making for historic buildings in the planning process has to be based on an understanding of not only the building's history and development, but also the current property market.

In any planning and development application, it is the obligation of the owner as applicant or their agent to provide the required documentation, including clear drawings indicating what is existing (historic) and what is a new alteration as well as information relating to the understanding and significance of the building. Negotiation between applicant and granting body is a part of the planning process. For the applicant/owner, establishing a dialogue with the local authority or representatives of the national heritage body early on will not only save time in the planning and decision-making process, but will also assist in developing proposals that are realistic and in the best interest of the historic building.

If the owner of a listed building is not carrying out the repair necessary to avoid further decay and deterioration of the building's fabric, then the State, possibly through the local municipality, may serve a notice to request that such repairs are carried out. Such enforcement is most effective where it is supported by some form of grants or assistance to the owners. In many developing countries, public and private sector money available for building conservation is severely limited and this undoubtedly impacts on how far legislation is enforced. Such enforcement is known as Urgent Works or Repair notice in England and the local authority has the legal right to carry out the works itself should the owner fail to respond to the notice, charging the costs to the owner. In exceptional cases, the State may take the decision to compulsorily purchase a historic property as the only means of ensuring its protection and adequate conservation. This is not legally possible in some countries. In the United States, the rights of a homeowner are primarily seen as being greater than those of the State, where the conservation of historic property is concerned.

Public participation

In our complex societies and increasingly global relationships, both the responsibility of guardianship and the right to decision making can become highly contested. Outside of the remit of legal ownership is the public or social ownership of a historic property by a local community, ethnic group or society as a whole. Moreover, the World Heritage Convention has introduced the concept of 'common inheritance'. This can cause conflicts in decision making and at times cultural heritage and its future can become highly contested, especially when the question of who makes the decisions concerning its future, from the building owner, to the architect who designed it (in the case of more recent buildings), the users, local community whose environment it is a part of, the colonial power who built the building or the colonised nation who has inherited it, society at large or the international community. In practice, however, most planning decisions concern local communities alone.

The State's decisions on cultural heritage are directly and indirectly informed by public opinion. Public opinion often plays an important part in what is listed and is an indication of what is being valued. New information technologies and the world wide web have made information held on monuments and historic buildings more readily available and accessible to the public through on-line databases, thus increasing public awareness and understanding of the built heritage. Media companies are also introducing more interactive forms of television programmes in which viewers are invited to take a more participatory role by voting to decide design awards, which historic building receive funding for conservation or even to elect a building to be demolished.

Public consultation is an important part of the planning process, especially where the physical and social impact of a proposal will be most obviously felt by the public. Public consultation ranges from inviting comments on applications to specifically organised events that target certain groups. Much of this type of activity, however, favours an educated and articulate public and may result in the opinion of the vociferous minority. There are a wide range of social research tools and methods that can be utilised to gather opinion to understand what different groups value in their environment. Ensuring that different interests are fairly represented in the planning process is important

Figure 4.1   Making conservation more accessible to the public encourages public participation and support for conservation.

and will only work if the time and cost implications are adequately planned for.

STATUTORY PROTECTION

Legislation in each country is organised differently with regards to how the cultural heritage is categorised, which is also linked to the level of protection and the statutory responsibilities of the various decision makers. In some instances, this has significant impact on the level of protection they are afforded. For purposes of legislation, protection and management, the most common divisions occur as follows:

- archaeological sites and ancient monuments
- listed buildings
- historic areas and groups of buildings
- historic gardens and landscapes

The first three of these categories also concur with the three categories of sites, monuments and groups of buildings that are recommended in the Granada Convention.

Archaeological sites and ancient monuments

Ancient monuments and archaeological remains, some of which remain below ground, are often protected under separate legislation. In some places, designation of areas of archaeological significance is divided into several categories. Such categorisation allows for areas with potential for archaeological discoveries, as well as buffer zones around known areas of archaeological significance, to be given lower grade designation that will provide some level of protection.

In England and Wales, archaeological sites and ancient monuments are designated as Scheduled Ancient Monuments, for which Scheduled Ancient Monument Consent has to be granted by English Heritage or Cadw if a proposed development is likely to impact on a Scheduled Ancient Monument or its setting. It is generally the case that Scheduled Ancient Monuments are not occupied and occupied buildings are listed. It may be the case that parts of a site are scheduled while other parts are listed. For example, where a Cathedral is listed, underground archaeological remains or an adjoining ruined cloister will be scheduled.

Many historic towns in Europe and the Middle East are built upon layers of history going back to Roman times and sometimes even earlier. In the centres of some of these important historic cities archaeology is an inescapable part of the development process. For instance, the construction of underground transportation systems in Rome, Athens and Istanbul had to be preceded by a series of investigations and excavations in each case, some of which yielded significant discoveries.

While the most important sites are most likely to be under State ownership, others, including areas designated for their potential, may be in private ownership. This brings with it conflicts, not only in the limitations for development but also for limitations it might bring to agriculture. Modern agricultural practices such as machine ploughing can damage material remains beneath the surface. New field divisions and single crop agriculture also impact on the landscape and in some cases on the setting of a monument, site or settlement. Such concerns, however, are often difficult to regulate or manage.

Increasingly, it has become the responsibility of a developer to assist in the cost of archaeological investigation. The biggest cost to a developer, however, is a time cost as the investment cannot be realised. Nonetheless, while there is always the chance of making an unexpected discovery, in most cases local plans can indicate areas of archaeological importance or significance based on known evidence and historic maps and documentation. As a development progresses, a trial or full excavation may become necessary. It may also be possible to determine the archaeological potential of a site through the use of non-invasive investigation methods, as will be explained in Chapter 5. In England, Planning Policy Guidance 16: Archaeology and Planning, provides guidance on how archaeology should be treated in the planning process.

Archaeological evidence is best evaluated in the context in which it is found and the preference is to retain archaeological material in situ wherever

Figure 4.2   The ruins of this church in Spitalfields, London, have been preserved in situ and can be viewed through the glass screens at the side or by walking over the glass cover.

this is possible and not detrimental to the material and evidence it contains (Figure 4.2). Where archaeological sites and ancient monuments are open to the public, the emphasis will be on protection, research and interpretation. The Archaeological Survey of India, for example, is primarily responsible for archaeological research and the protection of sites and remains of national importance. In other places, early settlement sites are protected under nature protection legislation or linked to anthropological study.

Monuments and buildings of architectural and historic significance

Statutory protection or listing the State's intention to protect cultural heritage assets from developments that threaten their architectural, historic and cultural significance. Buildings are 'listed' for a number of reasons and guidance relating to their selection will differ in each country. The choice of listing is closely related to the cultural, architectural and historic significance of a building or place, as explained in Chapter 3. But there will also be a number of other criteria to consider, some of which will only be of local relevance. Buildings may be listed for their:

- architectural value, including use of materials, construction techniques or demonstration of craft skills

- historic value and relationship to a certain period in history
- role in a significant event in history or association with a person of historic significance
- contribution the building makes to a group of buildings
- rarity value

Buildings are often listed for more than one of these reasons, and the combination of several aspects might increase the overall significance of the building and therefore the listing category.

Listing, however, will always be value-driven. At times, therefore, it is difficult to ensure entirely consistent comparisons across a country. Most listing is carried out as a national programme often as an ongoing process with regular updates. In England, additions to the list can also be made on the basis of recommendations from local authorities. At times it will be necessary to list a building as a matter of urgency if its integrity is seen to be threatened by a proposed or anticipated development. It is also possible to remove buildings from a list or register or to downgrade their classification. This could be the outcome of an appeal and re-consideration of its values, or because over time the building has lost the features for which it was originally deemed significant. Similarly, if new evidence comes to light or other aspects of its value are considered the listing classification might be increased. If consent is granted for demolition behind the façade of a listed building, then the façade may not necessarily remain listed.

A list or inventory of monuments or places of architectural and/or historic significance often has a number of functions, from being an inventory and database, identifying buildings and places worthy of conservation to classify them according to type, grading them according to their importance. Different grades or categories on a list also assist in identifying priorities where conservation is concerned, as well as the level of change that is permissible. At the time of writing in England, buildings are listed in three categories. Grade I listed buildings are deemed to be of outstanding or exceptional interest, while Grade II* listed buildings are of special interest but are not outstanding in the same way. Grade I buildings make up 2% of the total of the listed building stock and Grade II* buildings 4%; the remainder are listed as Grade II. Alongside buildings, a number of structures such as lampposts, post boxes, telephone kiosks and historic ships are also included in the list. In some countries, such as Denmark for example, different categories are not distinguished and all listed buildings and monuments are treated as equal in terms of legislation. In addition to a nationally compiled list, there may also be local lists of properties identified as being of local significance. Local lists play a role in informing local plans and might also indicate priorities for listing at national level at a later date.

Where historic buildings are categorised on a register, the highest status and level of protection will be monuments of national importance. Many are likely to have been the concern of conservation professionals for centuries, and not surprisingly monument conservation has been the starting point from which most legislation and conservation theory has been developed. Rarity also plays an important part in the choice to list a property. The older the structure or site, the more likely it is to be listed. In England, for example, all buildings built before 1700 are considered for listing, whereas only examples of the highest quality of more recent buildings are listed.

In other places, significant dates, such as those relating to before or after colonisation, may be used to categorise monuments or to inform decisions on their inclusion or exclusion from a list. In Greece, for example, buildings and monuments dating from the Ottoman period are not given the same level of protection as earlier monuments from the Hellenistic and Byzantine periods. Similarly in Jordan, buildings built after 1700 are not covered by the Antiquities Law. To date, only a few countries have legislation that supports the listing and protection of buildings of the twentieth century. In England, a 30-year rule, with the occasional exception if a building is threatened by demolition, is applied. In Denmark, 50 years is taken as the time after which a building can qualify for listing. In the listing of twentieth century buildings, the selection of the best examples is still likely to be subjective and is often open to debate.

A register or database of listed buildings will consist of a description of the building, including its exterior and interior features and, most importantly, make reference to its significance. In England, listing applies to all parts of a building even if these details are not recorded in the listing description, including anything fixed to the building, but excluding plant and machinery. It also includes buildings or structures within the 'curtilage' of a listed building, i.e. land which is linked to the listed building and is associated with its use, e.g. a garden would be an integral part of a house. Ideally, the listing description will clearly indicate the boundaries of the building or place that is being listed, including any outside areas that are an integral part of the building's design or setting. This may not always be clearly defined and could have implications on land development, in which case it is likely to be contested. Defining and protecting land and buildings that make up the setting of a monument may require separate legislation and is not always possible. In France, approval needs to be sought for developments within a 500 metre radius that may impact on the cultural significance of a historic monument, urban area or landscape. In some Scandinavian countries on the other hand, there is not even legal protection available to regulate building heights next to historic buildings.

Designation is not an end in itself but the start of a process to manage the cultural heritage, which will include:

- conservation
- changes and alterations
- development and additions
- demolition

Consent will be required to carry out these works to buildings that have been placed on a statutory list. Each will be treated differently by the law and determined according to the significance of the building and the case that is presented. Applications will be considered on the level of change and how it will impact on the significance of a building. Listed building consent will be required for changes ranging from painting and cleaning of exterior façades to major new extensions.

Alteration or extension to a listed building will require listed building consent depending on how these changes will impact on the character of the listed building. The procedure and granting of consent will depend on the regulatory system in place. Whilst development is managed and controlled by local planning authorities, the granting of consent for historic buildings is likely to be controlled by heritage bodies at national, regional or local level, depending on the significance of the building and the system that is in place for the implementation of the legislation.

Owners and users of listed buildings require clear guidance on the limits of what they can do. Because every historic building and situation is different, it is not possible to cover every eventuality through guidance. Nonetheless, guidance such as Planning Policy Guidance: Planning and the Historic Environment (PPG15) in England (1994) is helpful in informing on the implications of the legislation. The UK system is based on case law which can make it more flexible compared to the more rigid, regulation-oriented systems seen in most other European countries. There will always be a number of ways in which legislation is interpreted and implemented, including loopholes or other legislation that might be in conflict with the concerns of cultural heritage.

In more specific cases management guidelines, prepared in conjunction with owners, users, the local planning authority and other statutory bodies, can be used as a tool to interpret planning legislation and provide guidance relevant to the specific qualities and requirements of a group or type of properties (see case study box on the Barbican Estate).

Case study: Barbican Estate Conservation Management Guidelines

The Barbican Estate in London, built in the 1950s, was listed Grade II in 2001 (Figure 4.3). In order to address the concerns of the residents of the Estate's 2,000+ flats on what level of changes and alterations would be permitted in their properties, Conservation Management Guidelines were introduced. The guidelines were developed in collaboration with the local planning authority, the City of London Corporation, in consultation with the residents. The guidelines aim to inform residents, reduce the pressure on the Estate Department and streamline listed building consent applications being submitted to the Council.

The easy-to-use guidelines use a colour-coded system of green, amber, red and black. Potential works to the Estate and individual properties are listed under each category with an accompanying explanation. Works that come under the green category are acceptable

Figure 4.3   The Grade II listed Barbican Estate in central London where Conservation Management Guidelines assist in maintaining the integrity of the design principles while still allowing residents to change and upgrade their properties. (Photograph by Simon Woodward.)

and do not require permission. Interventions listed under the amber category are likely to be permitted, but advice must be sought beforehand, while those in the red category are only likely to be permitted through an application made to the Council. Works listed under the black category, on the other hand, are unlikely to be permitted. For example, while residents are allowed to remove and modernise kitchen fittings, the painting of the exposed external concrete surface, a major characteristic effecting the integrity of the complex, is not permitted.

There are instances where alterations to the historic character is the only way in which a building can continue to be used. For example, one of the outstanding features of the architect Eero Saarinen's iconic TWA Terminal at JFK airport in New York is the uninterrupted free flowing space of the check-in area. However, the need to address larger passenger volumes and increased security concerns at airports has resulted in the space being divided up, altering the distinct way in which the space is experienced and modifying the architect's original intentions.

At times historic buildings, including some iconic structures, have been demolished despite being listed, simply because without a viable new use the cost of ongoing maintenance was too high. One high profile example that received considerable publicity was the demolition of the Dunlop Semtex factory in Brynmawr in Wales, after a viable new use could not be found for the building that had also fallen into disrepair following 20 years of neglect. Consent for demolition will take into account the poor condition of a building, structural instability, lack of a suitable new use, the financial burden of maintaining a building with no end use, but a decision to demolish will always remain the last resort. Where damage to a historic building is deliberate or interventions have not been authorised, then legal action may be taken against the owner or those causing the damage, depending on the situation. In cases where historic buildings are part of a larger regeneration area, then their demolition may be seen as the 'price to pay' for the benefit the larger scheme will bring to an area. However, developments should always consider historic buildings on a site and investigate ways in which they may be incorporated into the development scheme if at all possible. In many cases they can add character, identity and value to an area, and as a result to the development as well.

Area-based designation (conservation areas)

The built heritage in its urban context is a multi-layered, multi-dimensional and living heritage asset, contributing to townscape character. Area-based designation and protection can relate to the entirety of a historic city, town centres, market towns, small villages, residential areas, groups of buildings, suburbs, designed towns, suburbs and settlements (e.g. garden cities), parks, gardens and urban open spaces. The designation and protection of entire areas is a much more recent development than the listing of monuments and sites, as discussed in Chapter 2.

The older quarters of Prague, in the Czech Republic, for example, have been protected by state designation since 1950 while in England planning legislation relating to 'areas of special architectural or historic interest' first appeared in 1967. In many countries, there continues to be no specific legislation concerning the historic urban fabric or the designation of conservation areas. Most often the recognition, protection and enhancement of historic areas falls within the remit of town planning and remains dependent on the policies and designations of local plans.

Figure 4.4   Old city walls, like those of Tallinn in Estonia, often create a boundary around a protected inner area, isolating historically significant areas that fall immediately outside the walls.

Like listed buildings, urban conservation areas are designated for different reasons, including:

- character and architectural qualities
- a uniformity that relates to a single period
- diversity that brings together influences and changes over time and different styles and contrasting scales
- specifically designed estates, settlements and cities
- links to the landscape and natural setting

Conservation areas may not necessarily be 'attractive' and there will be other values for which areas are designated. The buildings, group value, layout, morphology, open spaces, green spaces, links between buildings and to the wider landscape, setting and topography may all contribute to the distinguishing qualities of a place. The chosen boundaries of a conservation area impact on development as well as how areas within and outside it are treated. In some cases, boundaries are clearly defined while in others this is less obvious. The existence of old city walls, for example, is not necessarily the means of defining a boundary, as the areas immediately surrounding the wall may also be of significance (Figure 4.4).

Figure 4.5   Road networks and traffic signage concentrated on the periphery of the historic core in Salisbury, England.

Where listing of historic structures is usually undertaken at national level and is the remit of central authorities, conservation areas and the protection of the urban heritage tends to be the remit of local government. In England, designated conservation areas are the responsibility of a local authority and as a process therefore not connected to the listing of historic buildings. In other countries, historic areas may simply be a category in listing, and in some cases may also be categorised according to their importance and value.

Designated conservation areas, as well as other areas of historic importance or interest, need to be safeguarded and enhanced through local plans. The designation of large areas as conservation or protection areas has an impact on city planning as a whole. If they are not integrated within city-wide planning and development policies, historic core areas can become isolated 'museum' zones as the city planning process works around them. This is especially an issue as boundaries of conservation areas are extended and they grow in number. Another planning concern may be the policies for transitory zones that link historic areas to the rest of the city. In cases where central areas are heavily pedestrianised, then the edges will play an important role in traffic and transportation networks and if not managed sensitively add to the sense of isolation of the historic core (Figure 4.5).

Not all buildings in a conservation area will be of a historic or architectural significance that will necessitate them being listed. Indeed, some may be of limited architectural quality. Nonetheless, the demolition and replacement of any building in a conservation area will need to be considered and requires consent. In England, consent must be sought for any demolition within a designated conservation area. Furthermore, developments in a conservation area must enhance and not detract from the inherent character of the area. This ranges from cleaning, painting and change of colour, replacement windows and the attachment of canopies, advertising and shop signs. Some of these will particularly be a concern when the building in question is located in a conservation area and contributes to its character, even if it is not listed itself. Conservation area status brings new responsibilities for property owners, including issues regarding general maintenance and repairs to a property. For instance, cheaper solutions that might involve uPVC glazing or the replacement of slate roof tiles with the artificial variety will generally not be acceptable. However, it is vital that such areas are not 'frozen in time' and that the local authority is proactive in promoting appropriate and sympathetic change and development.

Many historic core zones have become extensive retail areas and concern for maintaining the character of a place against commercial demands will also be regulated by local planning authorities. Local planning departments will generally control the display of advertisements and signs on buildings and in areas of historic importance they may also impose additional constraints or regulation on the size, style and character of signage, to ensure that it is in keeping with the character of the area (Figure 4.6).

Historic gardens and landscapes

Historic landscapes can be classified as:

- designed landscapes (gardens, parks, open spaces)
- functional, evolved landscapes (continuing or relic landscapes)
- associated landscapes (location for events, battlefields, sacred places)
- landscapes that are settings to monuments and ensembles (such as the setting of a hill top Cathedral town)

In many places, the protection of historic gardens and especially landscapes is often linked to environment and nature conservation legislation. City parks, on the other hand, can be protected as part of urban conservation areas. Historic landscape conservation also takes many forms and there is even less legislation to protect historic gardens and landscapes. In England, a register is kept of Historic Parks and Gardens of Special Interest, although this does not have the same statutory holding or obligation as listed buildings. The register includes designed gardens, municipal parks and cemeteries. A second register of Historic Landscapes focuses on man-made features of the countryside such as hedges, walls, land division, tracks and the like.

In conservation areas, trees and other natural features may be protected separately. In England, a Tree Preservation Order can be placed on an individual or group of trees. Some hedgerows are also protected through the

(a)

(b)

Figure 4.6 (a) Local authorities can regulate how even globally recognised and standardised corporate signage can be made to fit into historic character of a place as in Mexico City; (b) rather than detract from the historic character as it does in Bologna in Italy.

Figure 4.7 Originally planted as small ornamental trees, these trees now overshadow the elevation and have roots that are too close to the 18th century building.

planning system, especially if they are in an area designated as a Site of Special Scientific Interest. In the management of conservation areas trees, hedgerows and other planted areas must be considered as part of the character of the area. Nonetheless, nature cannot be treated like a building, and at times it is more appropriate to remove dead or diseased trees, ones that are in danger of falling over or where the roots are causing damage to a historic building (Figure 4.7).

Other forms of control and legislation that impact on historic buildings

Intervention in and changes to historic buildings, sites and areas are not exclusively determined or controlled by historic buildings legislation. There will be other forms of regulation or legislation that will also impact on how historic buildings are used and altered. These include:

- development legislation (planning, regeneration, tourism, development frameworks)
- building regulations and building codes (see Chapter 6)
- other regulations (disability, access, human rights)
- other planning legislation (advertising, signage, licensing)

There may also be local bylaws in place to preserve a setting that frames a historic monument or views to it from key locations. For example, in London, key views of St Paul's Cathedral are protected. In other places, such protection may only be possible through local town plans.

While some legislation will have direct implications for cultural heritage, such as policies concerning pollution control may specifically include a clause for the consideration of cultural heritage, other legislation may have serious adverse impacts on cultural heritage protection. In some places, certain planning exemptions may apply to areas designated for major regeneration projects or even tourism development. This may also affect the statutory protection of cultural heritage, with serious consequences for its loss or poor quality conservation.

Acts governing the level of public access to historic buildings may also be contentious, with the degree of access often depending on whether a building is privately or publicly owned and what its current use is. Private owners may not have any obligations to open their properties to the public unless they have been in receipt of public sector financial support. In France, private owners are encouraged to do so through tax incentives and in England, public access for a certain amount of time per year may be a condition of grant aid for repairs and conservation.

SUMMARY AND CONCLUSION

The protection and management of cultural heritage is regulated through legal and administrative systems that are put in place at national, regional and local levels. Table 4.1 summarises a typical scenario of how legislation and decision-making responsibilities for different aspects of the cultural heritage might be organised. As with any policy or law on protection and conservation, it is not only the legislation but the system through which it is operated that determines success for its effectiveness.

Key considerations where legislation and policy are concerned can be summarised as follows:

- Decisions concerning the protection, management and conservation of the built heritage are taken at a number of different levels and by different stakeholders with different interests, some of which may be in conflict with one another.
- With better understanding today of cultural heritage as an integral whole, there is an even greater need to ensure that the various agencies are working in collaboration.
- The planning process is the key determinant of the level of intervention permitted for a historic building. The availability of sufficient knowledge and understanding of the building and its context will assist in informed decision making.

This chapter concludes the first part of this book. The next part places the theory discussed thus far into context through conservation practice.

FURTHER READING AND SOURCES OF INFORMATION

Clark, K. (2001) *Informed Conservation*. London, English Heritage.
Delafons, J. (1997) *Politics and Preservation*. London, Spon Press.
Larkham, P.J. (1996) *Conservation and the City*. London, Routledge.
Mynors, C. (2006) *Listed Buildings, Conservation Areas and Monuments*, 4th edn. London, Sweet & Maxwell.
Pickard, R.D. (2000) *Policy and Law in Heritage Conservation*. London, Spon Press.
Richards, R. and Urquhart, M. (2003) *Conservation Planning*, 2nd edn. London, Planning Aid for London Publications.
Suddards, R.W. and Hargreaves, J.M. (1996) *Listed Buildings*, 3rd edn. Gloucester, Sweet & Maxwell Limited.
Walker, R. (1995) *The Cambridgeshire Guide to Historic Buildings Law*. Cambridge, Cambridgeshire County Council.

Web-based sources

Council of Europe: www.coe.int
English Heritage: www.english-heritage.org.uk
Landmark Trust: www.landmarktrust.org.uk
National Trust, England: www.nationaltrust.org.uk
The Civic Trust, England: www.civictrust.org.uk
UNESCO: www.unesco.org/culture

Part II

Principles into Practice

Chapter 5
Managing conservation

Chapter 3 defined conservation as the process of understanding, safeguarding, repairing, restoring and adapting historic property to maintain its cultural significance. Historic building conservation projects range from small repairs to major alterations and adaptation to accommodate new uses. For the architect and other members of the project team, a conservation project will follow a similar trajectory to a new design project, although there will also be notable differences and additional obligations for team members. Most notably there will be a longer pre-design phase, where investigations need to be undertaken and an understanding of the property achieved. There will also be a greater involvement of the project team once the project is on site, as unknown aspects of the building reveal themselves. Not only will the works need to be followed closely but decisions will also need to be made as the works progress.

This chapter looks at conservation as a process, starting from surveying and analysing information, making investigations into a building and identifying the causes of decay and conservation needs, project planning, financing, implementation and evaluation (Figure 5.1). The ongoing management of historic properties and estates is discussed in the final section.

SURVEY AND ANALYSIS

Understanding

Conservation depends on informed decisions. Understanding is an essential first step in approaching repair, conservation or alteration of a historic building. Understanding also enables the better allocation of resources for maintenance, management and conservation. Works carried out without sufficient information being available can cause more damage to a building or trigger problems later on. The level of information needed for each project or situation will differ and considerations have to be made depending on the

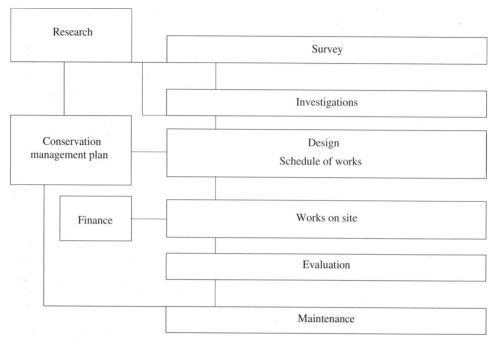

Figure 5.1 Diagram illustrating the conservation project process.

budget available for investigative work or research. Nonetheless, the more information that is available before a project commences, the greater the savings will be in the long term as fewer 'surprises' or unexpected changes will be encountered. Information and understanding also informs the planning approval process and will help guide decision makers.

Each building will have evolved and changed in different ways and it is through a detailed investigation of survey, observation and historic research that these changes can be understood. This should include any references to underground archaeological remains. For architects and engineers deciphering how a building is constructed and understanding the changes that have taken place is essential.

The level and depth of information needed will be dependent on the building and the type of intervention that is being proposed. Information regarding a historic building will be available in several forms. The first stage will involve research into the history, including context, of a building based on printed and archival sources; the second stage involves surveys that may range from a simple visual site survey to measured surveys, structural surveys and condition surveys through which defects and causes of decay are identified. Where sufficient information cannot be gained by these methods, then specialist investigative surveys will be commissioned using a variety of techniques depending on the information required.

Case study: Gloucester Blackfriars

Blackfriars in Gloucester is an example of a complex of buildings that have gone through significant changes over time, and where some evidence of each period is retained as a layer of the building fabric (Figure 5.2). Originally established as a Dominican friary in 1239, the church and other friary buildings were constructed throughout the thirteenth century. The Dissolution (1536–39) saw the passing of monastic properties to the crown and the buildings on the site were converted to new uses. The church was re-modelled to become Bell's Place, a substantial house, later divided into two, three-storey houses in the eighteenth century. All these later divisions were removed in the twentieth century to expose the single volume space to visitors. In two of the friary ranges, the thirteenth century roof beams survive, as does significant evidence of the monks' *scriptorium*, despite this wing having been used as a maltings and then as a mineral water factory up until the middle of the twentieth century. The west range was replaced by a row of houses in the early nineteenth century that still retain parts of the old cellars and several original walls. A simple brick building, used as an auto mechanics, 'Clutch Clinic' was inserted into the courtyard in the twentieth century. The site itself is known to be occupied since Roman times in the first century AD, followed by the Saxons in the eleventh and twelfth centuries, before becoming the site of a Norman castle in the early twelfth century. The friary was built directly on the site of the castle in the thirteenth century. Thus, any proposal to conserve, change or develop the site must consider this rich and complex layering of history that is manifested in the building's fabric, and its below ground archaeology.

Figure 5.2   Several of the many layers of history at Gloucester Blackfriars in England are still visible on the fabric.

Research

The significance of a historic building can only be defined on the basis of a thorough understanding of its historic development, which requires comprehensive research. Research will identify the various historic phases, developments and changes to the building that may not be immediately apparent from the fabric. Furthermore, contextual knowledge such as of the architectural styles and construction techniques of the period, as well as the social and economic conditions of the time will allow the values to be evaluated in these contexts.

Much of the information on a building or place of historic importance may already exist and it is often a case of knowing where to look. For example, the National Monuments Record (NMR), managed by English Heritage, holds a substantial archive of historic buildings in England. Additional information on other archives where information might be held can also be searched through the NMR website. Other information might be obtained from ownership documentation, leases, deeds, archives and local collections or historic records held by the owner. A local historical society might be of help whilst local authorities also often hold relevant information. The Royal Institute of British Architects (RIBA) maintains considerable information on historic buildings through its library, periodicals, photographic archive and drawings collection. Where the architect of a building is known, this might lead to further information on the building as well as other buildings designed by the same architect or practice that could provide helpful clues. More recent buildings may be published. In the UK, publications like Building have been in circulation since 1843 and provide short reviews and sometimes pictures or drawings of buildings. Initially, a building's owner might be able to provide information and a local authority may have copies of recent planning files. Ordnance Survey maps or historic maps, traveller accounts, postcards and photographs are also valuable sources of information.

In Europe, where a culture of writing, correspondence and diary keeping has been common practice for centuries, documents dating back to over a millennium can still occasionally be found and read as sources of information on historic buildings and their cultural contexts. For other cultures, oral traditions have been the key to passing on information and skills from one generation to another. Unfortunately, with rapid levels of development, urbanisation and the introduction of new technologies, this tradition is being lost, together with much of the information required for the care and maintenance of historic buildings and an understanding of their use and cultural context.

Library and archive-based research, essential for understanding the historic development of a building, should not be isolated from site visits and site surveys, through which research ambiguities can be addressed, information questioned and verified. Because a design has been submitted for planning consent or building regulations approval does not necessarily mean the building was built exactly to plan. A number of variations are likely to have taken place during construction that may not have been recorded and it is essential to check these in the field.

Survey

The purpose and scope of a survey may vary from a full measured survey, a detailed elevation study to schedule masonry conservation works to a five yearly (quinquennial) inspection to set out a forward maintenance programme.

It is advisable to visit the site before starting a survey to establish the size, access arrangements and limitations and to gather material such as records relating to previous repair works, and historic information. Plans, illustrations and photographs can be marked up as a way of recording information during a site survey. An understanding of the structure of the building and its construction will make it easier to identify the cause of defects or decay during the survey. A surveyor must ensure that the site and structure are safe before going on site, and to ensure that all health and safety precautions are in place. Where certain risks are involved, such as access by ladders, then there should be more than one person on site. Depending on the level of information required, binoculars are a safe and efficient way of surveying sections of elevation above ground floor level. Inspection is like detective work, searching for clues to the causes of decay, and the use of all five senses is regularly recommended. It is important that the survey is carried out in a methodological way to ensure every aspect that needs to be is covered.

If a measured survey does not exist, then one will need to be undertaken. Where measured surveys exist, they should be verified on site. Various methods are used in obtaining a measured survey of a building, ranging from hand surveys to the use of electronic and digital instruments. Image-based surveys include rectified photography which is two-dimensional, photogrammetry that will capture three dimensions and can be produced as a CAD (computer aided design) drawing and orthographs produced by digital photogrammetry. More recently, laser scanning has become available for the survey of buildings as well as objects. It is particularly useful in providing three-dimensional information through a 'point cloud' of measurements that enable a spatial representation of a building or object to be produced. For large sites, the use of Global Positioning Systems (GPS) as a common base is advantageous. With the increasing availability of GIS (Geographic Information Systems) various records and documentation can be correlated with plans and maps through an integrated database.

Despite these technological advances, the financial cost of a full measured survey will need to be considered, especially in places where a significant proportion of the cultural heritage has not as yet been documented and there is an imminent danger of loss. In such cases, low-cost solutions like semi-rectified photography using digital cameras and scales may be a cheap solution that will ensure a degree of recording is achieved. In many cases, surveys have proven to be the only record that survives of a historic building, placing the onus on the surveyor to be accurate, ensure the record is as complete as possible and the data is safely stored.

A condition survey should be presented as a report with accompanying drawings. The purpose of the survey or inspection needs to be stated and the method in which information is gathered and recorded clearly identified in the report as well as any access limitations. The report needs to be unambiguous and clear: it will not only be used to inform immediate concerns but should also be seen as valuable and accurate record for future use and reference. A structural survey will identify the structural form, loading and environmental factors and the condition of the various structural components. On complex structures, the use of real or computer-generated models may assist in understanding causes of defects, and especially structural movement.

Any inspection report will be based on established understanding of a building and will also inform the need for further investigations, such as potential structural problems that need to be investigated. A survey report is not a building works specification, but should contain advice on what works need to be undertaken and the urgency and priority of such works.

Specialist investigations

Levels of decay and causes of failure will initially be evaluated on the basis of the visual inspection. These problems might include level weathering, spalling, cracks, evidence of damp or staining on surfaces. Combined with knowledge of material behaviour, building structure and how the building is being used, the situation can be appraised and the cause of failure or decay recognised. However, visual clues may not always be sufficient and further specialist investigations may also need to be carried out depending on the nature of decay and the level of information that is being sought. These will range from testing the moisture content of materials, taking samples for testing or opening up areas to gain better access to a substructure.

At times, archaeological excavations will assist in providing an understanding of a site. If an excavation is carried out then all findings need to be clearly recorded and any remaining evidence carefully protected. Aerial photography and satellite images can also assist in identifying underground structures. Ground conditions can be investigated using geotechnical survey techniques including impulse radar or destructive techniques like sample pits or boreholes.

Several other techniques that assist in research include dendochronology, which identifies the age of timber and may assist in identifying the date or phasing of works. Paint analysis using microscope techniques will provide evidence of various decorative schemes over time and can play an important role in informing a colour scheme (see Chapter 7).

Wherever possible, the use of non-destructive survey techniques will limit damage or intervention with the historic fabric and cause less disruption to the building's users. Some non-destructive survey techniques that are currently available are:

- Radiography uses X-rays and works best through plaster or wood to determine structures or fixtures within.
- Impulse radar uses pulses of radio energy transmitted into a solid material or the ground to determine thickness, cracking, voids and any fixtures within.
- Microwave analysis measures the variations in the energy that is reflected back, and can indicate hidden details and faults.
- Thermography is generally used to measure temperature variations and heat loss in buildings but can also detect subsurface details such as a timber frame or a blocked up opening. A similar technology, infrared photography, uses infrared film to determine inconsistencies.
- Magnetometry identifies metal objects embedded in non-magnetic material.
- CCTV is used to inspect drains.
- A borescope with a fibre optic light source can be probed into cavities in brickwork or timber structures.
- Though marginally invasive, micro-drilling into timber helps to measure the level of internal decay.

Some information will only come to light when a project is on site. It is therefore paramount that no material is removed from the site prior to approval, thus allowing it to be inspected and recorded if necessary. Some unexpected clues may be found in this material, such as a piece of moulding or a fragment of an earlier colour scheme.

Information management

Documentation serves many purposes. Information relating to a historic building is not simply a record of the past, it is also a contribution to the general field of knowledge of a certain type of construction. Maintaining accurate and consistent records during works and opening up is essential as these records will provide important information relating to the historic building and the various changes and alterations that have been undertaken. Records will also provide information on the construction and condition of parts of the building that will no longer be exposed once the works are completed.

Records are best held by the building owner and should be passed onto new owners; copies should, where possible, be kept off site. In all instances, they should be easy to access. For buildings of outstanding significance and national importance records should also be deposited in a national archive that will enable them to be accessible to future researchers and conservation teams. Documentation in the public domain may provide valuable information that will assist another, later conservation project or enable value judgements to be made in a townscape character appraisal. In times of disaster or loss, such as fire, war, earthquakes, such documentation can be the only evidence that remains of a building. Where demolition and loss are inevitable, then documentation is the only way in which the information relating to the building and its construction is maintained.

National or local archives are an invaluable source in research of historic buildings. It is paramount that any new material that comes to light, from historic sources or through building investigations, are properly recorded and deposited in an archive for future use. Archives themselves need to be able to provide the best possible conditions and adequate space for the documents deposited in them, be well managed and easily accessible to their users. The storage of paper and photographic records require different environmental conditions for their storage and may therefore need to be stored separately. The digitalisation of records is a useful way of making them more accessible through the web, but remains a costly option to set up.

MAKING AND EVALUATING PROPOSALS

Conservation management plans

Conservation management plans are a valuable tool in informing conservation and design proposals for historic buildings and places. A conservation management plan will guide decision making during the planning process as well as inform the day-to-day maintenance of the building. Increasingly, local authorities and funding agencies require a conservation management plan in support of a planning or grant application. The aim of a conservation management plan is to identify the significance of a building or place of historic and/or architectural importance and develop management policies that will safeguard and enhance its significance and values. Conservation management plans vary in length and focus depending on the building or site and the scale of threats that are identified. The basic process of preparing a conservation management plan is illustrated in Table 5.1. The process involves three distinct steps:

1 Understanding the historic property/place and defining significance
2 Identifying threats to the significance and integrity of the historic property/place and responding with management policies
3 An action plan for implementation, evaluation of options and feasibility studies (Table 5.1)

Table 5.1   The conservation management plan process.

Understanding	Background		
	Significance	Values	
Analysis	Threats and vulnerabilities		Impact assessment
	Management policies		
Implementation	Action plan	Financing	
	Review		Update plan

Understanding

The first stage in preparing a conservation management plan involves the gathering of sufficient information to make an evaluation and objective analysis of the cultural significance of the building or site. This is often presented as a statement of significance. Any gaps in information that will require more in-depth research at a later time will also need to be identified. This will be followed by a list of the values that are recognised for the site. A list of typical values associated with the historic environment can be found in Chapter 3. Some values are derived from academic study or scientific research, while others are values attached to a place by its various user communities.

Analysis

The next stage is to evaluate the current situation and propose management policies that will protect and enhance the significance of the place. In order to make these proposals it is important to identify and evaluate threats to the significance and values, and the vulnerabilities of the historic property or site. This assessment must not be limited to the strict boundaries of a site as some threats or impacts on the site will be from outside. Opportunities for better management, change and development should also be considered at this time.

Policies and recommendations set out in a conservation management plan will provide guidance on conservation and safeguarding the significance. These management policies will respond to identified threats, and are intended to provide guidance for conservation works that are to be undertaken, development at or in the vicinity as well as the general care and maintenance of the property. All development and conservation proposals must be evaluated against these policies.

Policies that a conservation management plan will typically include will relate to:

1 The approach to the retention and safeguarding of the historic fabric and interiors;
2 The level and extent of change that would be permissible, including conditions of removal;
3 Guidelines and conditions for new interventions;
4 Access;
5 The preferred location for service routes;
6 The management, implementation and review of the conservation management plan.

Implementation

Depending on the situation, a conservation management plan may also include an implementation or action plan, including an assessment of cost implications and potential sources of funding. The action plan will indicate responsibilities for actions and a timeframe for implementation. Once agreed by all parties concerned a conservation management plan will become operational and should be consulted and referred to for any works undertaken at the historic property, as well as by other developments that may impact on it.

An important part of the process is the involvement of stakeholders and consultations to ensure that those trusted with the care and management of a property, statutory bodies as well as the users, are consulted on the proposed conservation policies and are supportive of them. Conservation and management planning must be seen as a process in which decisions are made as the plan is developed. Evaluations should be undertaken at the end of each step to assess whether revision, further research or analysis will need to be considered. A conservation management plan is not a static document and should be regularly evaluated and revised.

Impact assessment

An impact assessment considers the impact of new interventions and developments on historic buildings, settlements and landscapes against the significances identified for these places. Initially, it will inform the design process and recommend changes and alterations that will reduce adverse impacts. Where this is no longer possible, then it is important to identify the mitigation measures that will need to be undertaken. Where a large-scale development or change is being proposed, then an impact assessment may need to be undertaken to measure the impacts of a conservation or development project. A full environmental impact assessment (EIA) is an integrated process bringing together archaeological, historic buildings, townscape, ecological, traffic and other impacts together, allowing mitigation measures to be considered integrally rather than as separate concerns.

Development taking place in a wide radius can impact on the historic environment. In Mumbai in India, for example, pollutants from chemical factories on the mainland impact on the environment of the World Heritage Site caves on Elephanta Island, located some 9 km offshore. The impact of proposed high-rise buildings on the character and significance of the World Heritage Sites of Cologne, Germany, Vienna, Austria and the Tower of London in England have recently been contested. Development and change, nonetheless, is part of urban growth and the development process is important to maintain vibrant cities. Delicate balances often need to be sought in order to maintain the economic and social dynamic without detracting from the inherent value of the historic environment to the character of a place.

The methodology for an EIA is well established, and has also been adapted for historic buildings and areas. The first step is to identify the baseline conditions, depending on the scope of the study, for:

- above and below ground archaeology
- buildings of historic and/or architectural significance
- other structures or buildings on or around the site of architectural or historic merit
- setting and townscape character

Table 5.2 Measuring impacts.

Nature or impact	Level of impact	Duration	Reversibility	Permanency
Positive	Major	Short term	Reversible	Permanent
Negative	Moderate	Medium term	Irreversible	Temporary
	Minor	Long term		
	Neutral			

Once the baseline conditions have been established then the impacts can be identified and measured. Impacts are measured on the basis of being direct or indirect (knock-on) effects. Table 5.2 lists the various ways in which impacts are measured. Not all impacts are negative and detract from the historic environment. Some may also be positive in that they enhance the significance, such as the removal of a poor quality later addition for example. Some impacts may only occur during construction, but they also need to be considered and mitigated against to avoid any long-term damage resulting from the action. This might include vibration caused by pile foundations at an adjacent site, or damage to surface mouldings from scaffolding.

Various threats will have different levels of impact on a historic building or environment, depending on the level of intervention and the significance of the building or place. The criteria used to measure the level of impacts is summarised in Table 5.2. In an EIA, numeric values are assigned to the impacts, often ranging between +3 for a major positive impact and −3 for a major negative impact (Table 5.3).

Before a building project, conservation or otherwise, can commence, any adverse environmental impacts that it may impose on the natural habitat will

Table 5.3 Environmental Impact Scale (adapted from Morris and Therivel, 2001).

Impact

Major positive impact – Impacts are large in scale and/or to a building or structure of high historic or architectural significance (global/national) and/or its setting
Moderate positive impact – Impacts are moderate in scale and/or to a building or structure of historic or architectural significance (national/regional) and/or its setting
Minor positive impact – Impacts are small in scale (localised) and/or to a building or structure of limited historic or architectural significance (local) and/or its setting
Neutral – Negligible change is expected to occur as a result of considering the impact
Minor negative impact – Impacts are small in scale (localised) and/or to a building or structure of limited historic or architectural significance (local) and/or its setting
Moderate negative impact – Impacts are moderate in scale and/or to a building or structure of historic or architectural significance (national/regional) and/or its setting
Major negative impact – Impacts are large in scale and/or to a building or structure of high historic or architectural significance (global/national) and/or its setting

also need to be considered. This may include bats that that are inhabitants of a roof space or great crested newts that reside in nearby ponds and could be affected by building debris entering the water system at certain times of the year. In England, surveys carried out by Natural England will identify any environmental impacts that may arise while carrying out a conservation project or as a result of it. A consultant will also be able to advise on mitigation measures, such as timing of a project to ensure that it does not coincide with breeding periods. At times, conflicts will arise between the interests of cultural heritage conservation and nature conservation. In one London example, a medieval moat that to archaeologists may reveal significant information of the medieval occupation of the site, is today full of vegetation that is considered by nature conservationists as a rare area of biodiversity in an otherwise built-up area.

PROJECT IMPLEMENTATION

Financing conservation

The success of conservation or heritage-led regeneration depends on the ability of project champions to tap into sources of funding, whether it is small sums needed to support a conservation project or major sums required in regeneration grants to kick-start an urban renewal programme. Although the historic environment adds value and brings economic benefits to an area, funds still need to be secured for the conservation of individual buildings. As the remit of architectural heritage expands, the budgets available to support or grant aid conservation are shrinking. Even State parties looking after monuments of national importance have to raise external funds for major projects and the owners of ancestral properties are known to sell off valuable antiques to pay for major repair works. Several of England's major churches and cathedrals have turned to charging visitors entrance fees to be able to meet increasing maintenance bills and diminishing sources of financial support for such works.

Day to day maintenance work rarely attracts external funds and it is invariably 'invisible' and low profile. Most major donors like to support projects that have high visibility and that will produce good 'before' and 'after' images that demonstrate significant improvements when placed in brochures or annual reports. Likewise, major urban infrastructure projects are also less likely to attract funding. Most grants entail complex and time-consuming application processes that can be easily handled by larger and professional organisations, but can inadvertently discriminate against smaller and inexperienced applicants such as small charities raising money for a one-off project.

In the UK, the Heritage Lottery Fund (HLF) has emerged as a major grant giving body that is funded entirely by public subscription. Grants for historic buildings allocated through the HLF vary from significant contributions to major capital projects to small-scale grants for community schemes. In the case of large-scale projects, applicants must also be able to secure other sources of financing and demonstrate that future revenue will meet the cost of running and maintaining the buildings in question. The Architectural Heritage Fund (AHF), established in 1976, mainly provides grants for feasibility studies and small loans for charities undertaking building conservation. The AHF is also a good example of how a revolving fund can be used to help start of conservation projects.

A feasibility study carried out early on will identify costs and income streams and should evaluate the later maintenance costs for various options. There are times when a dedicated fundraiser will be required to identify sources of funding that might be available for a conservation project and to make the necessary applications.

Another way in which the State may support conservation is through tax relief on conservation spending, income tax, death duties, donations, value added tax (VAT), corporation taxes and local taxes. The way in which conservation projects are taxed also plays a significant role. Particularly where urban conservation is concerned, homeowners have often been encouraged to restore their properties through tax incentives and reductions. In some European countries, including in the UK, a situation prevails where VAT is charged for maintenance and repair contracts but not on alterations, consequently diminishing the incentive for conservation as it automatically increases the cost of works by a significant percentage (e.g. 17.5% in the UK).

A number of major international organisations such as UNESCO, ICOMOS, ICCROM and the Council of Europe, which provide conservation guidance, have been listed in Chapter 2. They are not, however, major grant providers for conservation projects. Even World Heritage Site status does not secure a grant, but indirectly a World Heritage Site is more likely to attract investment, grant funding or make gains through increased tourism activity. UNESCO funds are only available to World Heritage Sites in danger, and even that is a small sum that has to be spread out between an increasing number of sites. Only certain European Union (EU) Structural Funds can be used directly for conservation projects. Worldwide there are number of major funding organisations, amongst them the Aga Khan Foundation, the World Monument Fund and the Getty Conservation Institute. However, compared to the number of buildings in need of conservation and settlements in need of regeneration, the total sum is very small.

Once a conservation management plan is in place, necessary surveys and investigations have been undertaken, project proposals have been formulated, the impacts measured and necessary approvals obtained, then the project is ready to be implemented and 'go on site'. The consultant team will have been working together for some time at this stage, and will now be joined by the contractor's team.

Construction contracts for conservation

There are a number of ways in which a construction contract may be let. Either a single contractor will be appointed to oversee the project and employ subcontractors as necessary, or a number of subcontractors may be nominated in the contract. This may be essential in a conservation project to ensure specialist conservation work is being undertaken by suitably qualified teams. Often a careful balance needs to be sought between larger contractors' experience of managing a job and the smaller specialist contractors' skills in a specific area of conservation. In some cases, particularly in the case of a large estate of historic properties, the employer might directly employ and oversee the workforce. Today, there are a number of different ways in which building contracts are organised including management contracts or design and build. They may not all be appropriate for contracts concerning historic buildings. At times, it may be possible to adapt standard contracts for conservation works. This will depend on the legal framework for construction contracts in place in each country.

In most instances, it is the owner of a historic building who contracts conservation works to be undertaken and is therefore also responsible for appointing both the professional team and the contractor to carry out the works. There are exceptions, for example, when the State has intervened to carry out emergency repair works or is providing the funding for the project. In Spain, for example, State acts as the contractor for works on historic churches, although it does not own them.

Implementation on site

Unlike a new build project, there are often too many unknowns in a conservation project for everything to be determined beforehand. Further decisions and often more realistic decisions need to be made on site. Investigations will continue once a project is on site, when better access can be gained to certain parts. At times, fundamental changes to the approach and design may have to be made as a result of opening up on site. It may be appropriate to undertake limited opening up works during the design stage, where unknowns play a significant role in design and budget decisions that are being taken. Phasing projects may also be a way of avoiding too many unknowns from the start.

While at the research and project planning stages it is good practice to ensure that no compromises are made, as the situation on site may yet throw up a variety of unexpected situations where a balanced judgment will be called for and where a compromise may become unavoidable in order to reach an acceptable solution. Site-based decisions have to be taken within the restrictions of time, budget, client priorities, site working conditions and health and safety concerns. Conservation will always involve some element of compromise based on what is possible on site. For example, it may not be possible to gain sufficient access to a part of the building without causing damage to another part. In each case, a balance needs to be maintained amongst historic accuracy, the best interest of the building and its long-term stability as well as the needs of the users.

In conservation, the professional time needed on site is much greater than on a new design project, especially as questions are likely to arise as parts of the building are exposed and the quality of the original construction becomes apparent. Depending on the experience of the contractor, some conservation work may also need to be regularly inspected, or methodologies adjusted according to site conditions. Furthermore, regular inspection means that any historic details that are being revealed can be observed, acted upon and recorded.

While the individual sub-contractors carrying out some of the specialist conservation work will have skilled operatives on site who are aware of the historic significance, for other members of the team the project may be the first time they have worked on a historic building. It is important that the whole teams are aware of cultural significance of the historic building and the implications for site conduct and the treatment of materials on site. Scaffolding teams brought in as subcontractors are often least aware of historic significance and the erection and dismantling of temporary structures can be a point when damage is caused to the building fabric.

Traditionally, buildings were constructed and maintained in seasonal cycles. Most construction work would be undertaken in the spring and early summer, allowing time for the materials to dry out. External renders or plasters might be applied a year later, once a wall had sufficiently dried out. Maintenance measures would also be undertaken cyclically and seasonally, before or after a wet season for example. In many parts of the Mediterranean, it is still traditional in villages for houses and storerooms to be limewashed in the spring, since this is also a means of keeping insects out of rooms where the harvest will be stored.

In modern construction management, timescales are very different and are regulated by financing cycles more than the seasonal requirements of a building. Furthermore, there is a preference to undertake work on historic buildings that are open to the public outside the busy spring and summer tourist season. There can be serious consequences when using traditional materials for repair in the 'wrong' season and under construction management pressures for accelerated completion times. For instance, lime mortar should not be applied when temperatures are below 3°C; the mortar is also more likely to be subject to frost damage if it has not fully set and has not been adequately protected. Hastened work can lead to failure of an intervention only several years after completion.

MANAGING HISTORIC PROPERTIES

Maintenance planning

William Morris recommended building owners and users 'stave off decay through daily care'. The Venice Charter states: '*It is essential to the conservation of monuments that they are maintained on a permanent basis*' (Article 4) and the Burra Charter: '*The aim of conservation is to retain or recover the cultural significance of a place and must include provision for its security, maintenance and its future*' (Article 1). Lack of maintenance accelerates decay.

It makes environmental and economic sense to conserve and reuse buildings, yet there is a common belief that old buildings need continuous maintenance where new ones do not. This is not true, and the maintenance-free building remains a myth. All buildings need to be regularly maintained and the maintenance and replacement of some contemporary materials can be much more costly than ongoing repairs to traditional materials. Each building is made up from many different materials and while one material or component may require little maintenance, another, such as a joint sealant or a surface coating, will require regular renewal.

In some cases, building maintenance managers are responsible for a large estate or portfolio of property. Where the Oxford University Estate department looks after a stock of buildings ranging in age, with the listing of more twentieth century buildings the maintenance challenge on estates such as the London Zoo can be very different. For others a historic building may be a one-off on a large estate of largely less significant buildings. For a Defence Ministry, for example, a historic building may be an exception amongst a large portfolio of utilitarian buildings that include standardised office accommodation, residential blocks and aircraft hangars.

The day-to-day maintenance of a historic building will be undertaken or overseen by a number of people and contractors, not all of whom will have experience or expertise in building conservation. Establishing a set of guidelines beforehand is an invaluable way of ensuring that those commissioning, overseeing and undertaking the work are doing so in the best interest of the building. Regular testing and overview of services is also part of maintenance work, especially of those that could be a fire or flood hazard.

Preventative maintenance avoids later costly repairs and unnecessary change to the historic building fabric. An overgrown branch from a nearby tree can push a gutter up, causing the rainwater to penetrate into the brickwork and eventually damage an entire wall of a building. Similarly, blocked downpipes and gutters filled with leaves not only retain water and damp on a building they also help provide ideal conditions for new growth (Figure 5.3). Regular clearing of gutters can save a costly roof repair and internal plastering job later on.

Programmed or cyclical maintenance establishes the periodic cycle for which maintenance should be carried out and informs the annual budget. In the long term, this will be a substantial saving on repair costs if no maintenance is undertaken. Quinquennial inspections, undertaken once every five years, will ensure a building is regularly surveyed and allows for maintenance and repair works to be planned, budgets to be allocated and funds to be secured. If there is a conservation management plan, this can also be reviewed during the quinquennial inspection and updated. All proposed works should be planned and carried out in accordance with the conservation management plan. Only professionals with suitable experience and knowledge of historic

Figure 5.3 The lack of maintenance on this unused warehouse in Liverpool, England, has resulted in its being known as the Buddleia building.

buildings should be involved in specifying and managing maintenance works, repairs or proposals for alterations.

In the Netherlands, Monument Watch (Monumentenwacht), was started in 1973 as a service to owners with less experience in the care of historic buildings. The Monumentanwacht team carries out an annual or two yearly survey of the building and will advise on immediate- and medium-term works that should be considered. It is then up to the owner or administrator to appoint professionals to specify and a contractor to undertake necessary works. Similar initiatives have also been started in Belgium and the UK.

Risk assessment and preparedness

Like any building, historic structures are subject to a variety of risks from natural hazards such as high winds, freak storms and flooding to fires or deliberate vandalism. A risk assessment will identify areas of risk that need to be addressed in the protection and maintenance of a building. The probability of any given risk also needs to be assessed. Where solar radiation or fluctuations in temperature are an annual or continuing occurrence, high level flooding may be infrequent, but a risk nonetheless. Climate change is also now a recognised threat to historic buildings, and its impacts will need to be considered when long-term risks are being assessed.

In Italy, a GIS initiative that combines data on pollution levels, climate, earthquake zones, tourism flows and population density with the location cultural heritage sites, is being used to monitor impacts on cultural heritage. The system can also maintain records on the condition of the buildings alongside data from various research organisations and government departments.

Access management

In Chapter 2, the widening remit of cultural heritage was identified and the range of values that can be associated with the historic environment were discussed in Chapter 3. There is now an understanding of cultural heritage that it is of relevance to society at large and not only to those who are trusted with its care. Social and intellectual access are the remit of those concerned with the interpretation and management of cultural heritage, whereas physical measures that aid access to and within a building are the responsibility of building owners and their design often the remit of architects. In most historic buildings, some of the building's spaces will present limitations for access for people with mobility difficulties, especially changing levels and uneven surfaces. The balance between preservation and access is often seen as a difficult compromise. However, access involves community and for conservation to be successful and acceptable then the community need to feel part of the process.

Access is not confined to providing level access to those in wheelchairs, but finding ways to accommodate those with different disabilities including the blind, partially sighted, deaf and those with learning difficulties. Other considerations should include those with babies and young children and the safety of children. In developed countries, where populations are aging, provision for the elderly need to be considered in buildings and in the urban realm. Access strategies should also consider city-wide policies for better links between the city centre and surrounding area for example.

Physical access needs to consider both horizontal and vertical movement. One of the visually most obvious challenges is often a raised front entrance. Where ramps are often seen as being obtrusive to historic frontages, there are examples of sensitive approaches. The solution might not be in the building, but in making adjustments to the surrounding outside areas to provide level access, such as lowering or gradually inclining landscape features to provide

Figure 5.4 The central courtyard of the Royal Academy in London has been set on an incline to assist with easy access and to reduce the length of the ramp needed around the steps, allowing it to be neatly incorporated into the historic entrance.

entry (Figure 5.4). For vertical circulation, lifts are beneficial for people in wheelchairs, those with young children, the elderly and others with limited mobility. The location of a lift is therefore as important as the presence of the lift itself, if getting to and from it is going to involve a long walk.

In accommodating these various user needs, concerns may be voiced regarding the loss of historic fabric, loss of integrity of a historic building and aesthetic concerns, for example increasing contrast in paving materials that may not be seen to 'blend' in as well as another material (Figure 5.5). An access statement, which in England is a building regulations requirement, will explain how various access issues will be addressed in a building. There is no single right way, especially in the case of adapting existing and historic buildings. An access plan does however ensure that an overall strategic view is taken on how the building is used, rather than carrying out a series of reactionary adjustments. A review of a circulation pattern, for example, may make it easier for all to access and eliminate the need for a separate 'disabled' access. One of the simplest interventions is to remove obvious trip hazards.

The aim must be to enable reasonable, clearly signalled and dignified access for people whether they are visiting, working in, using or living in a historic building. It is important that solutions both respect people and respect the significance of the historic building.

Figure 5.5 Marking out the side of a step in white at Canterbury Cathedral in England, may not be the most elegant treatment of a historic interior, but it will prevent a considerable number of trips on the uneven surfaces.

SUMMARY AND CONCLUSION

The care, management, conservation and adaptation of historic buildings, like any design project, requires a methodological approach and a clear understanding of the process and the requirements of a historic building by all concerned. In summary:

- A conservation management plan will detail the significance and the values of the historic property as well as the conservation principles and policies that should be followed in its care, repair and conservation.
- Imaginative solutions can also be found for the financing of conservation projects whether it is through local grants, major donor agencies or private finance; or a combination of various sources.
- The use of appropriate investigative approaches at the project planning stage will ensure the project is properly informed and unnecessary surprises, delays and additional costs are minimised at the project implementation stage.
- Conservation projects should be approached with a clearly established philosophy, based on the understanding of the building and the needs of the end users, and communicated to all members of the team.

- On site, conservation will always require some element of compromise, but this should be the outcome of balance judgement relating to what is possible without jeopardising the cultural significance of the building.
- The care of historic buildings should not be seen as one-off conservation projects, but an ongoing process of maintenance, adaptation and evolution.

FURTHER READING AND SOURCES OF INFORMATION

Burman, P., Pickard, R. and Taylor, S. (eds) (1995) *The Economics of Architectural Conservation*. York, Institute of Advanced Architectural Studies.

Chanter, B. and Swallow, P. (1996) *Building Maintenance and Management*. Oxford, Blackwell Science.

Foster, L. (1997) *Access to the Historic Environment: Meeting the Needs of Disabled People*. Shaftesbury, Donhead.

Lichfield, N. (1988) *Economics in Urban Conservation*. Cambridge, Cambridge University Press.

Morris, P. and Therivel, R. (eds) (2001) *Methods of Environmental Impact Assessment*, 2nd edn. London and New York, Spon Press.

Sample Kerr, J. (1996) *The Conservation Plan*, 4th edn. Sydney, National Trust of Australia (NSW).

Swallow, P. *et al.* (2004) *Measurement and Recording of Historic Buildings*, 2nd edn. Shaftesbury, Dorset, Donhead.

Watt, D.S. (1999) *Building Pathology: Principles & Practice*. Oxford, Blackwell Science.

Watt, D. and Swallow, P. (1996) *Surveying Historic Buildings*. Shaftesbury, Dorset, Donhead.

Web-based sources

Architectural Heritage Fund: www.ahfund.org.uk

Heritage Lottery Fund for guidance on preparing conservation management plans: www.hlf.org.uk

National Monuments Record for England, accessed through: www.english-heritage.org.uk

Royal Commission on the Ancient and Historical Monuments of Scotland: www.rcahms.org.uk

Royal Commission on the Ancient and Historical Monuments of Wales: www.rcahmw.org.uk

Royal Institute of British Architects for access to the British Architectural Library, archive, photographs collection and the RIBA Drawings Collection online catalogue: www.architecture.com

The Aga Khan Trust for Culture: www.akdn.org/agency/aktc.html

UK Government National Archives: www.nationalarchives.gov.uk

World Monuments Fund: www.wmf.org

Chapter 6

Causes of decay, environmental services and structures in conservation

The structure, fabric, services and contents of historic buildings can be affected by decay. Before a structural and environmental conservation strategy can be determined, the reasons for material decay need to be understood in the context of how a building works. Overall, the knowledge required prior to making decisions on the conservation of structures and materials involves an understanding of the history and development of the building, as well as of its structure, construction and material components. Changes in internal conditions will also impact on the performance of materials and structure. Most importantly, it is vital to recognise that the various components, materials and services that make up a building are part of an integral whole and therefore must be able to work and function together. Approaches to repair and change, from major structural strengthening to a small patch repair of a material must be based on the principle that the new material, infill or addition must be able to behave in the same way as the whole. The principles set out in Chapter 3 also relate to the conservation of building structures and the accommodation of environmental services in historic buildings.

This chapter is in three sections. The first section looks at the causes of decay, illustrating that these are not all external forces, but may also be the outcome of how a building is being used or misused. The second section looks at ways in which modern-day services are accommodated in historic buildings as they are adapted to contemporary use and to meet current regulations and standards. The final section addresses issues of structural repair. It is impossible to separate building structures from the materials from which they are built, and more specific details of material conservation are covered in Chapter 7.

CAUSES OF DECAY

Identifying the presence of and recognising the causes of decay and failure in building structures and materials are an essential first step in any conservation project. Decay affects all buildings, whether old or new. All building materials and structures have a limited lifespan and will need to be repaired or replaced at some time in the life of a building. Changes to the use of a building and alterations over time may place undue pressure on the structure and thus material decay and subsequent loss of strength can be the cause of structural failure. All materials decay as part of the normal weathering process. A building surviving since antiquity is naturally going to be weathered and some components might be decayed and probably lost or replaced over time. Damage is when the structure is no longer performing to designed capacity due to decay, destruction or loss of components. The rate of decay is linked to the quality of the material, original workmanship, quality of design and detailing, level of maintenance or the way in which a building is being used.

Recognising the causes of decay will also inform maintenance programmes and the evaluation of risks discussed in Chapter 5. The causes of decay can be grouped according to their cause from various factors as being:

- Climatic
- Biological and botanical
- Natural disasters
- Human beings

Climatic causes

Traditionally, buildings were built from locally available materials and in forms and structures that best provided protection from local climatic conditions. Small windows seen in the Himalayas or in central African mud huts were not only a reflection of the small spans that could be achieved due to the lack of timber, but also a means of combating heat loss or excessive solar gain.

Solar radiation causes the break down of protective coatings like varnish or paint on external surfaces, the loss of which will expose timber window sills, for example, to rain and therefore water penetration. Sunlight or solar gain also causes fading of internal and external surfaces and fabrics in particular.

Water in any form causes or accelerates decay (Figure 6.1). Rot, insect and fungal attack in timber is more likely to occur when timber is damp. Externally, rain will streak, mark or stain a surface over time. Where the building is not sufficiently secure, or where elements of the building envelope are missing or decayed, then water will enter the building and start to damage the structure. The most vulnerable points are often the junctions where two materials meet, and around rainwater goods where water is specifically directed and collected (Figure 6.2). Water penetration around a gutter will impact on the timber wall plate or truss ends sitting in the external wall, causing them to rot and if untreated reduce their structural strength and capacity. Rainwater directed into an open gulley or broken drain can soak into the surrounding soil, leading to foundations being undermined. Frost damage is caused when materials that have absorbed moisture are then exposed to

Figure 6.1   Water penetration has caused the timber projection and frame to rot, gradually exposing the material beneath to the elements as well.

freezing conditions, causing the material to spall or break off. This is common in underfired brickwork or where the outer skin of the brick is damaged.

Rising damp is the result of water moving up through the building fabric as a result of capillary action, and is more likely to occur in porous stones or earth structures. The causes of rising damp can be varied. Although most traditional building techniques omitted a damp-proof course, it is often changes in ground conditions or ground levels that cause damp penetration and movement in the building fabric. Hardening of external surfaces around a building, for example, results in water accumulating against the base of a building rather than draining away naturally. If there are dissolved salts in a porous material, evaporation of the solvent will result in efflorescence, the deposit of crystals on the surface. This is frequently seen in new brickwork as the mortar releases moisture into the bricks as it dries out. Efflorescence in older buildings is more likely to be an indication of water penetration.

Different building materials react to temperature fluctuations differently, by expanding and shrinking. Often the movement is minimal and is naturally accommodated by the building's structure. Where temperatures fluctuate significantly between day and night or seasonally, then thermal movement of materials is likely to be more significant and may be the cause of failure of a material, especially where two different materials meet. Wind gradually weathers building surfaces, more so when particles such as sand are carried by winds against soft materials such as earth buildings. In seaside locations, salt carried by spray and wind rapidly corrodes metals. In hot and humid

Figure 6.2 A failed movement joint at Berlin's 1936 Olympic village is the cause of water penetration and further deterioration of the concrete.

climates, climatic factors often create environments that are ideal for insect infestation.

Biological and botanical causes

Biological causes of damage include insects, animals, fungi and plants. Birds, animals and vermin can also cause damage in buildings by dislocating materials or actively gnawing through them. Trees in close proximity to buildings can cause damage to foundations and walls with their roots, and differential settlement caused by adjacent trees is a common problem. Unstable overhanging branches, on the other hand, are liable to fall on buildings in heavy storms.

Insects mainly attack timber and include wood boring insects and beetles. Termites are found in warmer climates. In timber, the most common causes of decay are caused by fungal and insect attack. Fungal attacks are classified as being wet or dry rot. In dealing with fungal attacks, unless the cause of damp can be eradicated and better ventilation provided, it will not be possible to end the attack. Most moulds can be treated with fungicidal washes.

Cracks, blocked gutters and roof tiles where moss growth is occurring will provide ideal conditions for plant growth, especially from seeds carried by winds (Figure 6.3). As they grow plants can block rainwater goods, which can eventually cause substantial structural damage. There are varying views on the damage that may be caused by climbers, such as ivy on historic brick or stone walls. Ivy roots tend to grow into mortar joints and cause cracking in the masonry as they become woody. On external surfaces, lichens may react with stone and change its porosity, eventually causing surface damage. Lichens can also speed up the decay of mortar and of some surfaces. Although lichens will not damage timber, in large amounts they create moisture conditions suitable for fungal growth. Biocides are often used to control plant growth on buildings, but may have further reaching side effects on the environment.

Natural disasters

Throughout history, earthquakes and other natural disasters have led to the collapse not only of buildings, but even entire cities and civilisations. One

Figure 6.3   Through lack of maintenance, plant growth is beginning to take hold on the roof of this hamam (bath house) in Bergama, Turkey.

of the predicted outcomes of climate change is an increase in occurrence of freak weather conditions.

Any building in a seismic zone is prone to earthquake damage. Although many historic buildings were indeed designed and built to survive earthquakes, their condition and any later changes is often a defining factor in their survival. In some regions of the world, simpler vernacular buildings were not intended to survive the forces of a major earthquake but could be easily rebuilt. The Asian Tsunami of 2004 brought about devastation of a great magnitude as tons of water surged over coastal settlements. Besides the devastating human loss and suffering, damage included cultural property as well as the communities and their cultural traditions. Volcanoes also continue to be a threat in parts of the world, damaging everything in their wake. Because of their unmatched fertile soils, volcanoes have always been attractive for settlement despite the threat of eruption, Pompeii being perhaps the most famous historic example.

The flooding of Venice in 1966 proved to be a wake-up call for its conservation, although the design of various flood barriers are still being contested some 40 years on. More recent floods in central Europe have exposed the vulnerability of some of Europe's significant buildings and the valuable collections they contain to flooding. In some cases, the later insertion of damp proofing only hindered efforts to dry out buildings after the waters had receded. The storm and floods of 2005 in New Orleans that caused damage to much of the city's historic quarter and suburbs have also brought about proposals for rebuilding that threaten the character and integrity of areas surrounding and supporting the historic core. While conservation efforts focus on the 'tourist' historic quarter, many of the surrounding areas are now seen as prime development sites for new buildings that will also leave the central area as an isolated entertainment district that is no longer connected to the local population that give it so much of its character. Across the world, changing agricultural patterns, deforestation and mass urbanisation all play a part in land erosion and in the case of heavy rains devastating landslides.

Lightning striking and causing a major blaze in an historic building is not uncommon, especially for churches with high steeples. In 1984, the roof of the south transept of York Minster was hit by lightning resulting in a fire and significant damage to the entire transept roof. Alongside burning, fires cause smoke damage in buildings while the large quantities of water needed to contain and put out a fire can leave the fabric saturated. Even where no apparent structural damage is seen, the structural strength of a building will need to be reassessed following a fire. Buildings that remain exposed to the elements after a fire will continue to deteriorate if adequate protection is not put in place (Figure 6.4). Fires can be caused by humans intentionally in an arson attack or simply through negligence. Rebuilding after a fire also introduces much discussion on the level of reconstruction and rebuilding that should take place. For York Minster, one of the options considered was a new steel frame roof, but eventually a timber structure was chosen. The case study

Figure 6.4   A house totally gutted by a mortar bomb in Dubrovnik's old town in Croatia. The loss of the roof and exposure of the interior to the elements adds to the damage.

of rebuilding Uppark discussed in Chapter 3 highlights different challenges in post-fire reconstruction and conservation.

Human beings

Human beings it seems not only build and use buildings, but also cause significant damage to them. Most of this damage is simply due to day-to-day use, wear and tear, followed by lack of maintenance and poor building management. Inappropriate repairs and poorly considered alterations can also be the cause of serious or irreversible damage. Other man-made causes of decay are pollution, vibration and intentional damage to historic property through acts of vandalism, terrorism or war.

Urban densification, changes in water extraction or agricultural practices will change the water table and consequently impact on the foundations of a building or unexcavated archaeological material. Materials reach equilibrium with their surroundings; it is only when conditions are suddenly altered that a material that has survived in a certain condition for centuries starts to decay or disintegrate.

Atmospheric pollution has contributed considerably to the deterioration of buildings since the industrial era. Air pollution and acid rain react with the surface of materials, and more so if the material is porous (Figure 6.5). Traffic not only contributes to atmospheric pollution but can also be the cause of vibration that impacts on a building's structure. Vibration caused by pile

Figure 6.5   Atmospheric pollution is seriously threatening the rare inscription on the walls of the Augustus Temple in Ankara. (Photograph by Ufuk Serin.)

driving adjacent to a historic structure can be a threat to the integrity of the building's structure. In Athens, the flight path over the Acropolis was altered so that the monuments would no longer be exposed to vibrations caused by low-flying aircraft.

The poor management of building services such as heating or air conditioning systems can cause damage to historic fabric, as can the poor installation of plant, pipes and cables. Negligence can lead to leaks and the use of inappropriate cleaning practices can cause permanent damage to surfaces. Neglect of regular maintenance practices such as clearing gutters and keeping gulleys clear of debris can result in much costlier repair works later on.

Indoor microclimates also affect materials. Air conditioning, for example, can be damaging to the building fabric, while visitors release moisture that can impact particularly on art works or delicate decorative finishes. The extensive use of heating and different ventilation patterns today means that more moisture can become trapped in a building. Condensation caused by internal conditions is often a far greater cause of damp than rising damp caused by external conditions.

Historic buildings, and especially those with attached symbolic values, continue to be a target in wars, for terrorist attacks and deliberate vandalism (see Chapter 2). The impulse pressure produced by an explosion causes material and structural damage to buildings. While glass and other brittle or lightweight elements are likely to be directly damaged, the differential movement of each element in an explosion may cause cracking and structural damage to the structure itself.

On a more mundane level, there seems to be a basic desire among human beings to make their mark in the world by writing their names on the walls of well-known monuments. Although the evidence of similar activity by seventeenth century monks is now regarded as being of historic importance, the same is not true for 'Peter + Carla 2004' scribbled in felt tip in the narrow access corridor up to the dome of St Paul's Cathedral, the only place that is not monitored. Buildings of historic, cultural or religious significance also become the target of vandalism for increased impact. Graffiti and especially spray paints that are used can cause serious damage to historic building surfaces, especially where the surface is too fragile to withstand heavy and repeated cleaning or is porous and therefore easily absorbs the paint (Figure 6.6).

Tourists are also frequently blamed for the erosion of historic surfaces. Since it is the most significant historic buildings that attract most tourists, the consequence of the damage is even greater. Damage caused by high visitor numbers includes the erosion of surfaces, especially thresholds and stair treads, knocking against delicate surfaces, especially when a place is crowded, and alterations to the microclimate and environmental conditions of a place caused by human breath or micro-organisms inadvertently carried in on clothing. Whilst stiletto heels have long been blamed for puncturing holes in soft surface materials like timber, small particles of grit carried into buildings in rubber-soled shoes can also damage historic floors.

Figure 6.6 Graffiti is not only an eyesore, but can also damage historic building surfaces when repeat applications have to be continually cleaned off.

MANAGING ENVIRONMENTAL CONDITIONS IN HISTORIC BUILDINGS

Alongside repairing and maintaining the building fabric, conservation is also about adapting a building to new uses, contemporary ways of living and working. In a short period of time, significant changes have taken place in human demands for levels of heating and cooling in buildings as well as for the levels of sanitation that are now incorporated into buildings as standard (Figure 6.7). The relatively upmarket Midland Grand Hotel at St Pancras Station completed in 1876 contained only seven bathrooms for the 300 guest rooms. With the addition of an extension, the building is now being converted into a 244 bedroom five-star hotel and 67 high-specification and highly-serviced loft apartments. It has become almost impossible to maintain some historic hospital buildings for clinical use due to difficulties of accommodating the level of equipment and servicing as well as hygienic surfaces that have become a standard requirement in hospitals.

Introducing adequate levels of servicing is one of the major challenges of adapting historic buildings to contemporary use. This includes everything from lighting, power, heating, cooling, sanitation, drainage, access and egress. In addition, there will be the need to concur with current building regulations

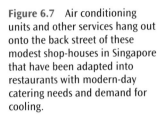

Figure 6.7 Air conditioning units and other services hang out onto the back street of these modest shop-houses in Singapore that have been adapted into restaurants with modern-day catering needs and demand for cooling.

and other statutory obligations, and to fulfil at least to some degree new requirements for energy efficient performance. Most servicing requires a horizontal network of pipework and cabling and the insertion of vertical shafts to accommodate lifts, staircases, service runs and sufficient provision to access them safely for maintenance purposes. The amount of space required and the openings that will need to be made in the historic fabric can become substantial if they are not adequately planned and sensitively executed.

One of the reasons for redundancy of buildings or their incompatibility for their intended use has been their inability to respond to changing environmental and servicing needs. In the UK, the property boom and speculation in the 1950s and 1960s created a considerable stock of office accommodation that became redundant because they were not suitable for the newly computerised workspace that not only required space for extensive cable networks, but also considerable cooling equipment to keep temperatures under control. Since then computers have become more efficient, produce less heat and can communicate through wireless and laser technology, while air extraction systems have also become more efficient and also require much less space. Building services technologies are rapidly catching up and responding to the needs of buildings. Wireless and fibre optic technologies, in particular, are helping reduce destructive cable runs in historic buildings. Although often a convenient solution for historic buildings, some of these new technologies still remain prohibitively expensive for most projects.

While listed buildings or buildings of special architectural and historic interest may be considered as special cases and in some instances can be exempted from building regulations, they are increasingly expected to fulfil statutory obligations. Nonetheless, the fact that historic buildings were not constructed with the current standards in mind, verbatim compliance with the rule book is not always possible. Most decisions will require balanced judgement that protects the significance of the building while achieving the best possible response towards fulfilling statutory requirements. For example, it may not be physically possible to install insulation to current regulation standards in a historic roof space, but the installation of some insulation will help to reduce heat loss. Other regulations that need to be complied with include those concerning the heath and safety of the building's users, and more recently those relating to access (see Chapter 5).

Figure 6.8 Cables and wiring randomly attached to the elevation of a historic building can become unsightly.

Designing for and locating services in historic buildings

Like all members of the project team, building services engineers working on historic building projects have to be sensitive to the historic significance of the building fabric and must recognise that standard solutions may not always be appropriate. Each building is unique and will require a customised approach to servicing in accordance with the needs of the user and the constraints of the building. Understanding how a building was constructed and how the materials work will assist in choices and decisions made to best accommodate new services. It is not only where services are located but also how they perform that will impact on the fabric of historic buildings.

Traditional building techniques worked on different principles of movement and breathability than modern-day structures, and changes to air movement and moisture content can impact considerably on building performance. Unlike structural interventions that are mainly aimed at strengthening the building fabric, services are a component of a building that become redundant in a much shorter time and require complete renewal. Even with relatively new buildings that would have been fully serviced when they were built, it is likely that the systems installed have become redundant and are in need of replacement. Existing services, even if redundant, may be historically significant and contribute to the understanding of the building and how it was used in a certain period. In buildings of a more recent past, services might have been an integral part of the design or the innovative nature of the building. While the removal of these often redundant systems is not desirable, they can create a substantial maintenance burden if left in situ.

Services impact on the fabric of a historic building in a variety of ways, from small holes drilled in architraves and ceiling mouldings to major installations of heating or air-handling systems. Constant changes and upgrades impact on the historic fabric as systems become redundant in increasingly shorter time spans. Building services installed today may only have a lifespan of 10–15 years or even less and therefore decisions relating to the positioning of services through historic fabric need to be carefully considered. The aim should be to preserve the historic fabric and extend the usefulness of a historic building, while at the same time reducing energy use and ensuring what is being installed is energy efficient and environmentally aware. This includes seeking reversibility and reducing irreversible damage. Architects, engineers and the building's owners or future users will need to work together to find the most appropriate and efficient solution that will serve the needs of the user, cause the minimum of disturbance to the building fabric as well as remain within budget. When designing services for historic buildings, careful consideration should be given to how much of the proposed servicing is necessary so that over-specification is avoided. The use of surface-mounted or free-standing units will allow for them to be easily replaced or upgraded. Fixings for cable runs, lighting or other services that are made into mortar joints will avoid damaging stone or brickwork. Drainage systems may impact on underground archaeological remains under and beyond the building.

Heating and cooling

Expectations of comfort standards have changed considerably in recent years and are very different from what they were in the past. However, heating or cooling historic buildings to new levels of requirements can have an adverse effect on the building fabric and its contents. While it is necessary that certain comfort levels can be maintained for a building to be used or visited, the impact of heating and ventilation on historic building fabric can be significant. For example, while modern day congregations seek reasonable comfort standards in churches, the introduction of central heating can have a detrimental effect on painted surfaces and frescoes. Even more detrimental is cyclical heating or cooling that creates variations in temperature and relative humidity (RH). Although fireplaces historically only provided localised heating, regular use over the course of a winter ensured that chimneys stayed warm and that the building fabric was kept at a reasonably constant temperature.

The heating and cooling of an interior changes the RH, which must be maintained within certain limits to protect both the building and its contents. RH is the 'ratio of moisture in the air to the moisture it would contain if saturated at the same temperature and pressure'. RH is inversely proportional to temperature and an increase in temperature will reduce RH.

Heating and cooling systems often require substantial amounts of ducts that will need to be routed through the building fabric. This will impact on the building fabric as well as its structural performance. Some services can be run underneath existing floorboards between the joists and room can be found for vertical runs in various voids, old service or ventilation shafts, redundant flues, through roof voids, or in storage spaces, closets and service areas. They may not, however, all be easily accessible for maintenance purposes. This should also be considered when services are run behind historic material (e.g. timber panelling) that may be exposed during building works, but will become difficult and disruptive to access afterwards. Furthermore, effective means of ventilation will need to be maintained to avoid condensation, and could include measures such as maintaining airflow through chimney flues.

Case study: Geffrye Museum Almshouse, London

One of the biggest challenges in the restoration of an almshouse built in 1714 as part of the display at the Geffrye Museum was reconciling the service and environmental needs of a museum open to the public with the desire to present to visitors an authentic experience of Almshouse life in the eighteenth and nineteenth centuries. Working closely with the services engineer and a heating subcontractor an underfloor heating system was devised that could be safely placed under the historic floorboards (Figure 6.9), with vertical runs being contained in a back cupboard away from the public display. The heating system is not visible to the visiting public, but provides uniform and ambient heat in all the rooms. All electrical cabling was similarly 'hidden' and sockets placed under floorboards with access for staff only. Fire detectors, alarms and emergency lighting were placed over the entrance of each room reducing visibility to visitors while providing sufficient cover. While the period display rooms have no lighting other than candle or gaslights, modern lighting is provided on staircases and circulation routes to enable safe use of the building.

Figure 6.9 Geffrye Museum, London: central heating pipes and cables are placed between the joists under the floorboards, thus ensuring they are not visible to museum visitors but allowing the museum to be able to maintain stable internal conditions.

Improving environmental performance

It is not only new buildings, but also existing buildings that need to improve their energy performance, to contribute towards the reduction of carbon emissions. The reuse of historic buildings is a sustainable practice of utilising an already existing resource. However, heating and cooling historic buildings to current-day comfort standards can be the source of considerable fuel consumption. Increasing levels of insulation in walls, roof spaces and under floor are amongst some of the measures that are being commonly undertaken. However, in introducing insulation the way many of these buildings function has to be considered as well as the environmental credentials of the materials that are being used. In the case of cooling, challenges can be greater as older buildings may have lost access to some of the external sources of air flow as they become hemmed in by new developments.

Insulation and dry lining, most of it devised for modern building techniques, must be added with care, especially to avoid condensation and added moisture, because previous ways in which moisture moved around a building are blocked. The use of breathable membranes helps to avoid some of these problems. Use of natural insulation products such as sheep's wool might be appropriate. However, sometimes the thinner profiles that are being achieved by other products are easier to fit into the limited spaces available in historic buildings. Air conditioning increases humidity in interiors and the use of electric ceiling fans might be more appropriate and often also more authentic in historic buildings.

Modern comfort standards have also resulted in buildings being draught-proofed to reduce heat loss. However, traditional buildings depend on some of this air movement for controlling moisture. Replacing traditional windows with new uPVC windows often results in a sudden increase in internal condensation. Plastic windows are rarely appropriate or compatible with historic buildings and there are other measures that can be effective in reducing draughts and managing heat loss or gain. For example, sash windows can be fitted with draught proofing; while the use of blinds, shutters, UV-resistant varnish or films on internal face of glass can help in reducing ultra violet radiation (see also section on glass in Chapter 7).

Lighting historic buildings

It is hard to imagine that most historic buildings would have been designed for different sources and levels of lighting than those that we have become accustomed to today. Much depended on daylight, supplemented by sources of artificial lighting from candles and oil lamps (see Figure 3.1 of the Parthenon, an interior designed to be lit predominantly from the oculus at the top). Gaslighting was introduced in the nineteenth century and there are a few buildings and streetlights that continue to use gaslighting today; modern-day safety requirements, however, require that certain adjustments are made to these fittings.

In most instances, lighting requirements are for a new use, whether it is sensitive lighting for the display of paintings or the provision of adequate levels of lighting for office use. In interiors, if the building was built prior to electric lighting then lighting is not authentic, but a necessary part of modern usage and living. The choice of light fittings must consider the appropriateness of the setting, but must also fulfil requirements for lighting levels. In a number of high profile historic buildings, freestanding uplighters are used to provide additional lighting that is needed for special events or functions. This allows for flexibility and ensures that historic surfaces are not cluttered with light fittings.

As much as lighting is an integral part of design, it should also be seen as an integral part of conservation. Lighting can be used to accentuate features and interpret a building's interior or exterior, offering users and visitors another experience. There may be a desire, for example, to highlight the vaults of a cathedral using uplighting, but lighting levels must retain a subtlety that does not overpower the lower parts of the cathedral, after all in the original candlelit night-time interior the vaults would not have been visible.

In the case where a historic building is open to the public as a display, then the lighting requirement may be to be as close to the original as possible in order to convey a near authentic experience to visitors (see case study box on the Geffrye Museum). Museums and historic properties also have collections of paintings and objects in them that will require special levels of lighting and small spotlights that might be trained on certain objects. Increasingly, this type of lighting can be provided by fibre optics and the light source need not be from the ceiling, but can be concealed in more easily accessible parts of the building fabric or furnishings. Glass fibre optics are relatively smaller than standard electrical cabling and therefore discreet and easier to install in historic fabric with limited disruption. They also require less maintenance than conventional sources and the light itself can pick out features and reduced glare. In museum environments, they can, for instance, be fitted into historic light fittings such as chandeliers and used to re-create the same quality of light.

For external lighting, the objective must be to bring out the key features and textures of the building without causing light pollution (Figure 6.10). Lighting a special feature such as a tower or an entrance porch might seem desirable, but seen at night the feature may appear out of context and detached from the rest of the building. Another consideration is for the placing of the light fittings and the visual impact this has on the surrounding landscape or even the building itself. Fixing lights and cable runs can damage historic fabric, while obtrusive light fittings, internally or externally, will detract from the significance and integrity of a building or monument. While white coloured floodlighting has a tendency to wash out buildings, the use of warm light sources will bring out the features of a building and works especially well on brickwork. Colder sources of light might be more appropriate for uniform surfaces like concrete. Night-time lighting does not have to mimic the day-time appearance and features around a historic building might

(a)

(b)

Figure 6.10   (a) Night time lighting can accentuate different features of a historic structure; (b) than seen in daylight.

Figure 6.11 The light fittings on this column in Jerash in Jordan is both damaging to the historic fabric that has survived from antiquity as well as to the integrity of the monument.

also be highlighted by using lighting in landscape features, such as lighting up the canopy of a tree. Lighting effects should always be tested on site before a project is realised.

Fire protection

Many historic buildings were not designed to comply with current-day fire regulations. Tellingly, fires were a constant threat and building acts were often developed as a means of improving fire safety in cities. For instance, much of the evolution of warehouse design from the eighteenth to the nineteenth century was to reduce the risk of fire which was leading to substantial commercial losses. Fire continues to be a major risk to historic buildings and their contents. Furthermore, some historic buildings contain very valuable collections of international significance that also need to be protected.

Approaches to fire prevention need to be sensitive to the historic fabric and integrity of the building. Taking a fire safety engineering approach for the entire building rather than trying to make each member or element compliant with building regulations is more likely to produce an integrated approach that protects the significance of the building while still ensuring adequate fire safety coverage. This can be achieved in discussions with a fire officer who can advise whether alternative interventions will achieve the same level of safety.

Major historic buildings should have a fire safety statement as well as a plan of action in case of fire. A detailed fire risk assessment of a historic property will help recognise areas of potential danger and establish measures that can be put into place to reduce this. The age of some installations such as wiring or other electrical faults should also be regularly reviewed. Another important consideration is to ensure that fire tenders will be able to access the building in case of a fire, that there is a ready supply of water and that all access routes are kept clear at all times.

In many instances, historic buildings need to be upgraded to reduce the risk of fire spreading, protecting human life as well as the irreplaceable historic fabric. Despite the value attached to the historic fabric, the safety of those using a building must remain a priority. Fire safety concerns for any building include adequate means of escape and safe egress for anyone in the building; containing the spread of flames and smoke internally and on the external envelope. Structural elements supporting other structures above them will need to have a level of resistance to avoid immediate collapse in a fire.

Fire prevention measures in historic buildings include a range of interventions from upgrading components to the installation of fire detection systems (see also Figure 7.4, Queen Charlotte's Cottage). The type of fire safety measures that most often need to be incorporated include additional escape routes or partitioning within a building, all of which can have an impact on the fabric and character of historic buildings. When upgrading doors and windows to improve their fire resistance, the addition of intumescent strips or closers are appropriate in some cases but not in others and may need careful consideration. The application of such measures will need to consider the historic value of the building components and be undertaken with the overall fire safety of the building in view. Most importantly, service runs will need to be sealed particularly where they penetrate the building fabric. Some gaps can be filled with non-invasive materials such as fire pillows. Improving ventilation in a roof space will not only reduce the risk of fire but also ensure that fire and smoke go upwards rather than spread down into the building.

Fire detection systems will detect smoke, heat or flames and early detection of fire, particularly in high risk and inaccessible places such as roofs, can save historic buildings from substantial damage. These can now be radio linked, thus avoiding extensive wiring. It is essential that a sufficient level of detection is in place for areas where fire is less likely to be detected, such as in roof voids. Other requirements for public buildings include emergency lighting and illuminated exit and directional signs. Their installation without detracting from the historic integrity of an interior, while still providing protection to those using the building, will be another design consideration. Preventative measures like the installation of a sprinkler system, must consider the risk of water damage, but there are instances where this may be a feasible protective measure in fire protection.

STRUCTURAL CONSERVATION

A building's structure has three main functions: to provide strength, stiffness and stability. Strength is its ability carry loads, including the weight of the building as well as other applied loads; stiffness its ability to support the loads without undesirable movement; and stability the ability of the structure to stand up as a whole. In some cases, the whole may refer to more than an individual property, but maybe a whole terrace.

Structure often determines shape and sometimes also the aesthetic of a building. In conservation, priority has to be afforded to ensuring the structural stability of a building, in some cases with interventions that may conflict with the principles of minimum intervention and reversibility. Nonetheless, the basic principles and approaches outlined in Chapter 3 should form the basis of the conservation of structures and there exists sensitive and sympathetic approaches to most problems as well as good practice examples. If indeed the structure of a historic building has to be altered so much to accommodate a new use, then maybe the chosen use is not that appropriate.

Structural failure

Traditional building structures from grand monuments to modest vernacular dwellings are based on the building techniques of post and lintel construction or arches, vaults and domes. In masonry structures, it is often the geometry and shape that provides the stability more than the strength of the component parts, while vertical surfaces such as floor plates add rigidity. In traditionally built buildings, piers and columns are most likely to be masonry or timber. Metals were first used to join and fix components but following the industrial revolution iron and then steel started to be used as structural elements in buildings. Since the beginning of the twentieth century reinforced concrete has become a popular and relatively cheap means of construction.

New or additional loads on historic buildings will threaten their structural integrity. Most traditional materials work best under compression, wood being one of the few materials resisting both tension and compression. Metals will suffer from fatigue when they have been stressed beyond their elastic limit. In earthquakes, seismic waves produce both vertical and horizontal forces. Because buildings are designed to resist vertical forces, most earthquake damage is created by the horizontal forces, resulting in cracking and disconnection between structural members. The leaning Tower of Pisa may be the most famous example of the impact of soil settlement on a building, in the case of Pisa caused by differences in the base soil.

The visible signs of structural failure are cracks; crushing and buckling; and deformations including bending. Cracking is often caused by low resistance of masonry structures to tensile stresses and the settling of various materials in a given situation. An analysis of cracking patterns will establish the cause of the movements, if the movements are historic or ongoing and whether they are structurally significant. This analysis will inform the severity of the defect and the remedial measures that need to be taken. Most cracking will not pose a serious threat to the stability of a building, though should always be investigated and monitored. Crushing is far more dangerous and there is often little warning before failure occurs.

There are various devices to monitor cracks and measure movement. Even where a decision has been made not to intervene, it will be prudent to continue monitoring a structure for any significant movement. This can be achieved by placing small strain gauges on a building that will be used as locating points from which regular measurements can be taken. Alternatively, optical instruments or a theodolite can be used to measure vertical movement. Placing 'tell-tales' across cracks or an opening where movement may take place is a well-established method, but can be misleading as a crack in a glass tell-tale may be caused by other, climatic reasons. Monitoring should be an aid to, not a replacement for, understanding structural issues.

Understanding and analysing structures

The approach to the conservation of building structures is an integral part of building conservation and follows the principles of research, understanding and analysis prior to commencing works. The ICOMOS International Scientific Committee for the Analysis and Restoration of Structures of Architectural Heritage (ISCARSAH) has proposed a series of principles on the analysis and repair of historic structures, in which a proposed methodology is set out as: survey, test, diagnosis and treatment, report back. One of the greatest challenges facing engineers working on historic structures is maintaining a balance between preserving the material and structure and making sure it is stable and safe.

When investigating historic structures, assessments and calculations that would be applied to a new structure are not necessarily the best tools in yielding a complete understanding of how the structure is behaving. Firstly, buildings such as a medieval church were not originally built on the basis of calculations but on tried and tested methods and indeed trial and error. Therefore, to base a structural appraisal on calculations used for modern-day construction will not necessarily yield appropriate conservation solutions. Secondly, even considering the construction technique and the properties of the materials used there will still be an element of unknown from the way the structure has moved and settled over time. Sometimes even slight changes will make it behave differently. Thirdly, information needed for engineers to calculate the performance of a building may not be readily available. A timber frame structure might be concealed under plaster or brickwork or a weakened wall plate may not be immediately accessible. Even when the frame is exposed, it will not always be possible to know the condition of the joints. In concrete frame structures, unless accurate design information is available, then the reinforcement within the concrete will also be invisible. Thus, much will depend on opening up, investigative surveys as explained in Chapter 5 or carrying out various tests on the structure. And even when original design intentions are known, the reinforcements may not have been built as designed or they may have lost some of their strength due to corrosion.

Lessons of performance and repair can also be found in history, as most building techniques have been developed through trial and error, mistakes have resulted in collapse. The dome of the Hagia Sophia in Istanbul, for example, was fully or partially rebuilt four times before sufficient buttressing and an iron chain around the base of the dome secured it against earthquakes. Working with historic buildings requires some intuition, but where there is doubt then the opinion and advice of a structural engineer with an understanding of historic buildings should be sought.

The appraisal of a historic structure is a combination of engineering expertise, knowledge of historic construction techniques and professional judgement. Some consider the understanding of historic structures and their performance to be more of 'an art than a science' (Figure 6.12).

Figure 6.12   The structural repair and strengthening of the monastery wall in Valdigna, Spain, has been achieved by using reinforced concrete that has been coloured to match the original stone wall. This level of intervention, though probably structurally secure, must be carefully considered in terms of aesthetics, understanding and integrity of the whole.

Repair and conservation of historic structures

Structural repairs to a historic building will involve repair, reinforcement, replacement with a replica or replacement with a new structure. The aim is to ensure strength and stability of the structure and robustness. The choice of repair should not simply be based on cost but also on durability and long-term maintenance needs.

Repair and respect for the original fabric is the basis of structural repair, with reinforcement where necessary, but in ways that will not compromise the integrity of the building. In certain cases, it may be necessary to introduce a new structural system. Even when existing materials are maintained, this will require careful thought and ethical consideration. The future stability of York Minster was achieved by inserting new concrete foundations under the slim gothic pillars. The carefully engineered foundations do not in anyway impact on the interior space of the Minster, but can be viewed by those visiting the undercroft.

Foundations are often the starting point for structural repairs since without sufficiently stable foundations any work carried out above would only be superficial. Where the capacity of the existing timber piles are in doubt, underpinning or concrete piling are amongst the methods used to strengthen foundations. Remedial works to foundations are the most costly and if the foundation is sound and there is no evidence of ongoing movement, then works could be avoided and strengthening carried out elsewhere in the structure.

Many will question why a building that is standing and appears safe should need to be strengthened. Although a historic structure might be revealed as inadequate in light of present-day building regulations, if it is not going to be exposed to any more loading than it has been historically then this could be seen as acceptable. This will often be the case for domestic buildings. It is, however, prudent to ensure at this stage that there are no defects that might change the structural stability, such as rotting beam-ends or missing components. Sometimes, simple measures like effectively keeping water off a building or avoiding water logging around foundations will be sufficient in maintaining the structural integrity of a historic building. Where the loading needs to be increased, this should always be discussed to establish whether the changes are appropriate, and then an analytical model of the structure and testing may become necessary. Detailed survey drawings, and in some cases modelling, are also very important for engineers to understand how a building is performing.

In the case of cast and wrought iron framework structures, where corrosion is minimal and not seen to be reducing structural performance then it need only be removed and the ironwork painted with a protective coating to avoid future corrosion. If the iron member has fractured then it will have lost some of its strength; if the elements cannot be repaired then they will either have to be replaced with steel or supported with new materials. In reinforced concrete structures carbonation causes cracks and eventually compromises the ability of the concrete to protect the steel reinforcements from corroding. The treatment of reinforced concrete includes the removal of loose concrete and cutting back as necessary to establish the level and extent of corrosion. If the structure is deemed to be sound, then the corroded metal will be cleaned and treated before new concrete is cast back in. Where the damage is severe, however, new reinforcements may need to be added and this may alter the depth of the concrete casing. The conservation of reinforced concrete is discussed in more detail in Chapter 7.

In masonry structures, stitching with stainless still anchors is used to tie in delaminating walls, while grout is used to fill in voids and to consolidate loose rubble insides of walls. The use of Portland cement for repairs or grout is not desirable since it sets too hard, is not compatible with weaker and softer traditional mortars and it fails to move with the building.

Temporary works are used to provide access as part of a planned work and conservation programme to historic structures, to provide protection or to support a failing or unstable structure (Figure 6.13). However, a contractor's standard scaffold solution may not be appropriate to the structural conditions

Figure 6.13 Following earthquake damage, this church in Peubla, Mexico, is being supported by a temporary structure while conservation and structural strengthening works are carried out.

of a historic building. For instance, it is advisable not to fix scaffolding into the fabric of a historic building, to avoid loading sections of a wall unduly and any unnecessary damage to the building surface. Nor should scaffold trusses rest on a roof truss, unless it has been proven capable of taking the additional load. The involvement of a qualified engineer who is aware of the structure of the historic building will also significantly reduce the risk of damage to the historic building.

SUMMARY AND CONCLUSION

There are a wide range of causes of decay and damage of historic fabric, some of these are avoidable, others are not. However, even when decay and damage are unavoidable there are a ways of reducing the risk or their impact through forward planning, regular maintenance and appropriate environmental and structural interventions and fabric repair.

Some of the key considerations of this chapter can be summarised as follows:

- Any intervention towards adapting and upgrading the structural and environmental performance of a historic building requires a good knowledge of how a building works, its original construction, how it has been altered and repaired, how it has been used and how well it has been maintained.
- Although the adaptation of historic buildings to current-day regulations and environmental standards can present some serious challenges, these are also the opportunities for creative solutions.
- Building in flexibility in conservation and services will make it easier to accommodate future change.
- It is essential that historic buildings are viewed as an integrated system when repairs are being carried out on various components or materials.
- New materials in structural repairs will only be acceptable where they are compatible, appropriate and have been tested.
- A balance will need to be sought between maintaining as much original fabric as possible and addressing future maintenance needs (especially where access, opening up and scaffold costs need to be taken into consideration).

FURTHER READING AND SOURCES OF INFORMATION

Baer, N.S. and Snethlage, R. (eds) (1997) *Saving Our Architectural Heritage: The Conservation of Historic Stone Structures.* Chichester, John Wiley & Sons.

Charles, F.W.B. and Charles, M. (1984) *Conservation of Timber Buildings.* Cheltenham, Stanley Thomas Ltd.

Croci, G. (1998) *The Conservation and Structural Restoration of Architectural Heritage.* Southampton, Computational Mechanics Publications.

David, J. (2002) *Guide to Building Services for Historic Buildings.* London, CIBSE.

ICOMOS International Wood Committee (1999) *Principles for the Preservation of Historic Timber Structures*. Available at: www.icomos.org/iiwc

Institution of Civil Engineers (1989) *Conservation of Engineering Structures*. London, Thomas Telford.

Larsen, K.E. and Marstein, N. (2000) *Conservation of Historic Timber Structures*. Oxford, Butterworth Heinemann.

Levy, M. and Salvadori, M. (1992) *Why Buildings Fall Down: How Structures Fail*. New York, W.W. Norton.

Phillips, D. (1997) *Lighting Historic Buildings*. Oxford, Architectural Press.

Robson, P. (1991) *Structural Appraisal of Traditional Buildings*. Dorset, Donhead.

Robson, P. (1999) *Structural Repair of Traditional Buildings*. Dorset, Donhead.

Ross, P. (2002) *Appraisal and Repair of Timber Structures*. London, Thomas Telford.

Salvadori, M. (1990) *Why Buildings Stand Up: The Strength of Architecture*. New York, W.W. Norton.

Warren, J. (1999) *Conservation of Earth Structures*. Oxford, Butterworth-Heinemann.

Web-based sources

Building Research Establishment (BRE): www.bre.co.uk

Getty Conservation Newsletter: www.getty.edu/conservation

The Timber Research and Development Association (TRADA): www.trada.co.uk

Chapter 7
Conservation of materials

This chapter is concerned with the fabric of buildings: materials and their conservation. From the early shelters constructed by prehistoric man up until industrialisation, when new materials started to be manufactured and mass produced, the palette of materials available for construction was fairly limited and restricted to natural sources. Nowadays, architects and engineers have at their disposal a wide range of materials from which to construct buildings. Alongside materials commonly used in traditional construction, this chapter also addresses newer building materials, especially concrete and plastics common to the modern movement and more recent buildings.

Traditionally, buildings were made from natural materials and depended on their permeability and natural ventilation to maintain their integrity as a building. Modern construction techniques rely on different principles and materials to achieve the required performance standards and are guided by the building regulations that are in place (Figure 7.1b). In conservation, compatibility of new materials with the original is a key issue and one that might conflict at times with the practice of using different materials to indicate repairs or later interventions. Many of the new materials used today may have a very short lifespan. A case may arise where the original choice of material may not have been fit for the purpose or proven too susceptible to weathering in the given location. If the replacement of like with like is not sustainable, then a similar but more durable material might be considered. In some cases, careful and intuitive detailing might solve the problem. For example, extending lead capping on a parapet that will be invisible will provide better protection for the masonry beneath.

In the conservation of materials, there is no single correct technique or technology. Furthermore, repair will be closely linked to the building techniques specific to each region (Figure 7.1a). Materials used on buildings are closely linked to craft skills and traditions. The continuation of these skills is part of the conservation process and should be the first indication of how to approach conservation. Research continues into many areas of material conservation and repair and some methods or materials pioneered in the recent past have already proven to be inadequate and in some cases even detrimental

(a)

(b)

Figure 7.1   (a) The construction and maintenance of this village house in Tibet depends entirely on materials available in the immediate environment. (b) Most modern buildings such as these concrete structures are the result of globally developed technologies which necessitates a complex supply chain and specialized skills to construct and maintain.

to the performance of the original fabric. It is the role of the conservator to understand not only the physical but also the chemical characteristics of materials.

The focus of this chapter is commonly used building materials that constitute and contribute to the building as a whole. The conservation of interior details and finishes are touched upon, but this field is left for the skills of the specialist conservator and is covered in much better detail elsewhere. For each material, its common uses and application in traditional construction is given by way of an introduction. The causes of decay and failure are discussed and the principles and techniques of repair are given. The chapter is by no means exhaustive and the information provided is intended as being only of an introductory nature. The aim here is to provide guidelines and considerations for good practice and not to prescribe repair methods, which must always remain situation specific. The latter are covered in much greater detail in other publications, a selection of which is listed at the end of this chapter. In preparing specifications for conservation work, advice should always be sought from specialists and the most recent publications consulted.

EARTH STRUCTURES

Building in earth

Earth is one of the world's most commonly used building materials. Earth construction is based on the use of a locally available resource that, on redundancy, will either return to the ground from which it came or be reused in repair or rebuilding. Earth construction to this day remains one of the most environmentally sound building techniques, using material that has been excavated at the site and eliminating carbon production from production or transportation of the material. The use of earth is not restricted to low-rise domestic buildings. Some of the most outstanding structures in the world have been built out of earth, including the slender tower houses of Yemen and the Great Wall of China.

Earth construction takes many forms but its use is modified in each region based on climatic need and the materials that are commonly available. The main types of earth construction are:

- Dried mud bricks or blocks that are used in construction with mud or lime mortar
- Earth in semi-plastic form, constructed without the need for mortar
- Rammed earth made by compacting earth between two restraining surfaces (Figure 7.2)
- Earth infill panels in timber frame structures
- Earth compacted or mixed with lime to form a weak screed for flooring

As a material, earth has compressive strength but limited tensile stress. The cohesion of earth material is possible only if the construction is kept dry and in that sense it is a high maintenance material. It cannot be point loaded and care must therefore be taken to ensure that any restraining and supporting structures employed during conservation works are able to spread load across a surface.

In earth construction, cohesion is achieved by clay or, in the case of chalk, fine chalk dust. A clay-rich material has more cohesive strength. The porosity of unbaked earth means that it is prone to expand and shrink considerably on water intake and in drying out, respectively. Earth buildings traditionally combined materials (masonry, mortars and finishes) of similar porosity,

Figure 7.2   Chapel of Reconciliation, Bernauer Strasse, Berlin, uses the remnant material of the dynamited nineteenth-century church in its earth material base, thus retaining the memory of the former church on the same site. (Photographs by John Stevenson.)

Figure 7.3   An abandoned tower house in southern Saudi Arabia. Earth structures are highly vulnerable to the elements if they are not maintained.

strength and thermal properties to work together efficiently, which is why combinations with materials such as wood, metals or glass can be problematic. The addition of straw or other reinforcing natural fibres adds strength to earth structures, and may be part of the mix, or laid down in layers as a wall is being built up. A plinth of stone or boulders often protects the base of a wall against rising damp from the groundwater or rainwater splashing against it. In Britain most earth structures are constructed from clay lump, known as Cob or compacted chalk. Wattle and daub panels placed in timber frame buildings are common to most of northern Europe. Surfaces are usually rendered, often with lime renders and finished with a limewash.

Causes of decay and failure

The decay of earth is usually caused by water. Heavy rain can very rapidly endanger such a structure, while groundwater and rainwater splash is often the cause of damage to the lower sections of a wall. The structural failure of earth structures is often caused by water penetration, loss or failure of reinforcing timber components and vertical cracking (Figure 7.3). Thermal movement and ensuing expansion and contraction is another cause of vertical cracking in earth construction. By its very nature and porosity, an earth mix is an ideal environment for plant growth. Plant growth will either cause structural damage or, in the case of smaller growth within pores or crevices, the organic acids reduce the cohesive properties of clay.

Most traditional earth structures lack basic amenities and as they are modernised with new services, water leakage or change in humidity caused by an air conditioning system can accelerate material decay. The use of hard renders and cement mixes in combination with earth or for repair purposes also accelerates decay. Similarly, most modern paints that do not breathe are not appropriate for earth walls (see Figure 7.29).

Principles and techniques of repair

In conservation terms earth structures will need to be treated differently from masonry or timber structures, since:

- Earth buildings cannot sustain the passage of time; an earth wall that starts showing the effects of time is also rapidly decaying and exposing the inner core to weathering.
- Earth structures require continuous care and maintenance; renders and limewash coats will require regular renewal and this must be carried out using the same material.
- Unlike timber frame or masonry buildings, it is very difficult to move and reconstruct earth buildings.

Generally, the utilisation of traditional methods should be the first option in repair. However, repair techniques will vary from construction techniques since the same level of access is no longer possible, such as in rammed earth. The size of infill and its position will determine the repair technique and choice of materials. The structural repair of earth structures is complex, as infill clay is likely to shrink and not sufficiently bond with the substructure. Where a mix is being applied directly to a fill using shuttering, the material must be reasonably stiff and shrinkage minimized. The new material will be eased into the opening and rammed in to fill all voids. Clay tiles may be inserted into the back of the wall to carry the weight of the repair and to act as reinforcement, tying the repair back into the wall. In some cases fibreglass bars or metal mesh reinforcements are used to increase the bond between existing and new material in repairs. It is best not to insert new damp proof courses into earth structures, but to ensure there is sufficient drainage around them. Cement-based fillings are not compatible with earth and the repair of cracks should be undertaken using a clay-based grout.

It is possible to salvage material for reuse in earth construction so long as it has not been contaminated. However, individual mud bricks cannot be reused as they will have lost some of their strength and it will only be possible to use the original material as part of a new mix. Blocks can be produced and then cut to size to fill holes created naturally or where material has been cut out due to failure.

Figure 7.4   Cement-based mortars and renders are not compatible with soft materials.

In a wattle and daub or lath and daub building, new wattle panels may need to be inserted if the timber has been subject to insect attack and can no longer be maintained. If not, the area of damaged daub should be cut out, the wattle cleaned and lightly sprayed with water, and a repair mix that matches the original worked onto the wattle.

A render needs to move with the whole, and therefore must be of a mix and consistency that is able to do this. Wetting the existing surface in the traditional way will usually provide sufficient bond. Renders with mud contain a mixture of clay, silt, fine and coarse sands and up to 5% fibres such as straw and/or dung and need to mature for 24 hours before application. Shrinkage should be tested before the material is applied. Any cement-based render will harden and not move with the wall, often trapping moisture behind it and subsequently causing more damage (Figure 7.4).

THATCH

Thatching

The use of local grasses, palm fronds, straw and specially grown reeds in simple or sophisticated applications as a roofing material can be traced back to prehistoric times and some of the earliest forms of shelter. Thatch, being a relatively flimsy and vulnerable material, is not one that has survived well (see Figure 3.8 of a thatched roof exhibited in an open air museum in Tokyo).

Three types of thatch commonly used in England are long straw, water reed and combed wheat reed. Straw is the stem of cereal plants and a by-product of traditional agriculture. Water reed for thatching is cut from managed reed beds that are grown for the purpose. Combed wheat reed is actually a variety of straw that is prepared and fitted like water reed. Ecologically, thatch makes use of locally available materials and managed reed beds can also play an important role in wildlife conservation and in maintaining biodiversity.

Declining demand combined with changing agricultural practices means that locally available supplies of much thatching material are diminishing. For example, straw in the form that can be used for thatching is no longer an output of harvesting. The availability of cheap alternatives has also seen thatched roofs the world over, being replaced with materials such as corrugated iron, the conservation of which is discussed later on in this chapter.

Causes of decay

Most palms, grasses and bamboos are vulnerable to insect attack, whilst rodents will try to feed off grains left in the thatch. Ensuring water runs off a roof efficiently will avoid damp conditions. Thatch is also highly vulnerable to fire, especially in dry conditions and a fire in a thatched roof can also go undetected for some time. The firefighting approach is often to remove the thatch to ensure that parts do not continue smouldering. This not only is highly destructive, but also leaves the building vulnerable to the elements. Sprinkler systems have been installed in some high-profile thatched buildings, such as London's Globe Theatre. These will drench the roof at the first sign of a fire (Figure 7.5). Such preventative measures against fire, however, will be prohibitively costly for many buildings.

Principles and techniques of repair

Repair of straw or combed wheat reed is often a case of removing a loose and unsound layer of straw or reed down to a sound base and then fixing the new thatch to this layer. This practice not only preserves material in cost terms, but also ensures that evidence of earlier material and techniques are maintained for the benefit of archaeological research. In England, for example, some material closest to the rafters could be medieval. Rethatching a reed roof, on the other hand, often involves stripping the old reed back to the roof timbers. A thatched roof will require some level of renovation every 20–30 years.

Conservation of thatch, like earth, results in a new top layer obliterating evidence or patina of a previous period. Whereas in the case of earth the loss might have been a decorative detail, in thatch it is often the detailing around

Figure 7.5 A sprinkler system being fitted as a preventative measure to the thatched roof of Queen Charlotte's cottage, Kew Gardens, London.

eaves, windows and in particular the form of the ridge that can be altered. The choice of the 'right' style might prove problematic since photographic, historic and archaeological evidence may point to different types of material and ridge detailing at different times in a building's history. Covering straw roofs with a wire netting discourages birds and other rodents that might be attracted to remaining grain particles in the straw or for nesting material.

TIMBER

Building with timber

Along with earth, timber has been one of the earliest building materials available. Timber has traditionally been used for framed structures and for most roof structures. The size of locally available timber has often determined the size of the members and construction techniques employed. In places where timber was available in abundance, it was also used for entire buildings such as log cabins, as a cladding, for weatherboarding and for decorative features. Other uses of timber include the use of bark or sections of timber as roof shingles. These often have quite short lifespans and need to be replaced every 40 years or so. In Scandinavia, Nepal, China or Japan, major monuments were built entirely from timber. Timber is also used for window frames,

Figure 7.6 Timber is used both as a structural and decorative material in Nepal.

doors and extensively in interiors as panelling for walls, on ceilings and as floorboards (Figure 7.6).

Both the type of tree and place from which the timber member is cut will determine the strength and characteristics of building timber, especially in terms of susceptibility to fire or insect attack. A living tree typically holds moisture within its cells and once cut, timber will shrink and may warp, but with the addition of moisture timber is also liable to swell. The inner section of a tree is known as heartwood and the outer section of new growth as sapwood. Heartwood is more durable, has lower moisture content and is less susceptible to insect attack; there are, however, significant variations between different species. Timber is often classified as hardwoods and softwoods. Softwoods are coniferous species, grow fast and are the source of most commercial timber. The differentiation of hardwood or softwood is not directly linked to the strength of the timber and the quality of timber from different species within each group will also vary significantly. The strength of timber is dependent on the way forces act in a structure in relation to the direction of the grain. Where forces are in the direction of the grain, timber is strong in bending, compression and tension; but the same is not the case when

Figure 7.7   Timber frame building with infill panels in Safranbolu in Turkey's Black Sea region.

forces are acting perpendicular to the grain, in which case the compressive, tensile and shear strength of timber is reduced.

The timber frame of a building would historically most commonly be filled in with wattle and daub or bricks (Figures 7.7 and 7.8). Battens and plaster, tiles or weatherboarding are also common types of facing and enclosure, protecting the frame from the elements.

The main form of a timber frame for a roof is the triangular form of two rafters held together by a tie. Loads are transmitted in compression from one member to the other and the triangulation provides stability. Joints between the timber elements are therefore crucial to the performance of the structure. Timbers are joined by means of timber to timber (e.g. mortice and tenon, scarf, notched lap) or with metal fasteners or adhesives. Timber is a naturally durable material and the use of the right type and cut for the purpose and its adequate protections will give timber a very long life as a building material. Timber is nowadays machine cut as standard, and most significant spans are achieved by glue-laminated members. Metal fasteners, including nails and bolts, used to join timber members have been mass produced since the nineteenth century.

Figure 7.8 An exposed timber frame prior to restoration in Quedlinburg, Germany.

Causes of decay and failure

Since most timber frame structures depend on triangulation for stability, the loss or deterioration of an element may be the cause of structural instability and consequent movement in the building. Stresses in a structure are concentrated around joints and deterioration around joints often needs to be carefully monitored. A structural survey will identify the structural form, loading and environmental factors and the condition of the various structural components (see Chapter 6). Other defects, such as knots, wanes, distortion or splits, also weaken timber.

Some amount of deformation, such as deflection of a member, might occur and the implications of any deformation to the stability and the safety of the building will need to be assessed. Creep deflection over time will result in some visible deflection of members. This is difficult to reverse and as long as the slightly deformed shape is acceptable to the user, it can be left as it is.

Increased moisture content makes timber an ideal host environment for fungi, and timber needs to remain dry to perform its structural role. Where rot has been identified in timber, the first step must be to identify and eradicate the cause and then allow the timber to dry out. New timber is often supplied treated with preservatives, most effectively using a vacuum or pressure process. For existing timber, newly cut or exposed ends can be treated by brushing a preservative on the timber, but this is unlikely to penetrate deeply into the timber. The use of chemicals to treat attacks or to impregnate timber does, however, have environmental and health consequences.

Timber is also highly vulnerable to fire and history is full of examples of mass destruction brought on by blazes from the Great Fire of London (1666) to the many blazes that swept through Istanbul's tightly knit neighbourhoods of timber houses in the eighteenth and nineteenth centuries. In Finland, there is little remaining evidence of vernacular buildings much earlier than the eighteenth century, simply because they have all burnt down.

Principles and techniques of repair

In Japanese tradition, a timber structure is completely dismantled and rebuilt every 300–400 years with partial repairs being carried out in the intervening period. This is not necessarily wholesale renewal, but ensures that all defective materials are replaced and joints once again strengthened.

For Western practitioners, this is less likely to be the chosen method of conservation. Original material is embedded with valuable historic information, including the age, form of cutting and masons tool marks, as well as clues to changes that might have occurred, such as marks where laths or other fixings might have been. Replacement members will no longer contain evidence of that period. The ICOMOS International Wood Committee's Principles for the Preservation of Historic Timber Structures, adopted in 1999, advocates the use in repair of the same or similar types of timber, construction and jointing techniques that were used in the past for each particular case. However, where past details or methods have proven inadequate and are the cause of decay, these should then be remedied rather than repeated. Timber conservation, like all forms of material conservation, depends on the availability of carpenters with the knowledge and understanding of traditional construction methods and tools.

The aim of timber conservation must be, as in all forms of structural conservation, to make the existing building stable, not to expect it to perform as if it were a new building (Figure 7.9a). Typical interventions in a timber structure include replacement of a member, replacement of part of a member (often an end) or patch repairs. Repairs may depend entirely on timber replacement or may include the use of a metal bracket or strap or the use of an adhesive.

The replacement of part of a timber member is often carried out by cutting out a rotten or defective section. Cutting out rotten wood must ensure that all traces of fungal attack are removed with a reasonable safety margin to ensure no active attack remains in the timber and all new timber must be treated. Then a new section can be spliced in with what is known as a scarf repair or scarf joint (Figure 7.9b). This may take on a myriad of forms and may or may

(a)

(b)

Figure 7.9   (a) Timber frame building and infill panels being restored in Erfurt, Germany. (b) Detail of repair to a timber beam.

Figure 7.10 The recent restoration of this early timber frame building in Quedlinburg, Germany, involved jacking up the frame to repair the base.

not involve the use of metal fasteners. The key consideration, however, will be to ensure the repaired beam or member, which will not be as strong as the original single member, continues to act as one, especially in terms of bending moment and in retaining tensile strength. The use of reclaimed timber is often not recommended for repair purposes as it may not be possible to verify its structural properties.

Where it is not possible to repair a timber structure using in situ techniques, whole or partial dismantling might need to be undertaken. Timber frame structures are also capable of being dismantled and reconstructed at a new location, if such a need arises. Most replacements can, however, be undertaken by jacking up the building (Figure 7.10). The most likely member that will require replacement is the cill plate that is closest to the ground and absorbs the most moisture; this may also be an opportunity to improve the stone or brick foundations on which the timber frame sits.

In cases where the timber might be less visible, the use of metal reinforcements for repair is one way of strengthening the member while retaining most of the original material or a timber frame might also be reinforced by adding steel members. There are also structures that were originally built using timber and iron components. Particularly the early warehouses of the industrial era combined iron columns with timber beams (see Figure 7.24). These types of repairs are to a large extent reversible. Strengthening work might also be necessary to allow new loading requirements for a new use or for the addition of new services, heating system or insulation panels. It should be noted, however, that metal components are vulnerable to corrosion and, unlike timber, will fail when they reach high temperatures in the case of a fire.

The use of epoxy resin and polyesters is a more recent innovation in timber repair. It will bond to timber as well as steel and is most likely to be used to bond steel into timber. Resins do not, however, allow for reversibility of a repair. Generally the use of fillers in timbers is not a necessary measure to improve performance, but allows for the retention of more original material and is most often used for aesthetic purposes. The use of fillers on external timber or at joints, however, is preferably avoided.

Regarding other timber components that are used in buildings, timber windows can usually be repaired by replacing or piecing in new timbers in areas that have deteriorated. It is often not necessary to replace the whole frame but just the lower rail where rot is more likely to occur due to failing paintwork and water penetration. In interiors, most historic panel surfaces would have been painted to hide deficiencies in the timber as well as providing a coating that would have taken the daily wear and tear. Stripping timber will often reveal a less than desirable surface and is rarely historically accurate.

If well looked after, timber floorboards will survive relatively well with only small repairs. It is best to avoid cutting holes into the timber to route services, as the services are likely to become redundant long before the floorboards will need changing. It is more difficult at the present time to source replacement boards of the right width that have been sufficiently seasoned so as not to warp. Wooden floors would have traditionally been stained, painted or treated with substances like linseed oil and beeswax and then buffed with a soft cloth.

STONE

Building with stone

Stone is a natural material used in buildings in many forms, from rubble that is readily available to cut and carved stone. Its durability means that even ancient stone structures have survived to this day and a substantial amount of historic evidence of earlier periods is carried in stone (Figure 7.11). Stone continues to be a material of choice where prestige is concerned.

The type and durability of stone differs according to the way the particular stone was formed, its chemical and mineral composition and its texture. Igneous rocks are formed by volcanic activity and include granite and basalt. Sedimentary rocks are formed by accumulation and compaction of rock waste and include limestones and sandstones. Metamorphic rocks are formed through the alteration of igneous and sedimentary rocks by heat and pressure and include marble and slate. Artificial stone is a composite cast material of crushed stone, stone dust and Portland cement. Although it will give the appearance of a stone, its structural properties are more akin to precast concrete.

Figure 7.11 Detail of a wall from the Hellenistic period where the various marks show the setting details for the next row and the position for the iron cramps which would have been secured in place with molten lead.

Stone can be used as it is found in nature in applications like dry stone walling, in coursed or uncoursed rubble walls, or it can be cut and dressed as ashlar. The structural use of stone is possible only where the bending stress is low; thus, spans are often achieved through arches and vaults and stone lintels are used only for short spans and are often quite deep. Stone is also commonly used for flooring, paving and as a roofing material in the form of stone tiles or slates. Small cobble stones are used for road surfaces and have become associated with historic towns. Cut stone continues to be a signifier of prestige and expense, and it is not surprising that stone has been a much-copied material. Brick walls have been rendered to look like stone since the Renaissance (Figure 7.28) and the technique of scagliola developed in the sixteenth century is a plaster finish imitating fine marbles in interiors (see section on Finishes later in this chapter).

Figure 7.12 Even within the same wall different stones can weather at different rates.

Causes of decay and deterioration

Substantial international research continues into the causes of decay, means of measuring rates of deterioration, cleaning and repair of stone. Degradation of stone surfaces can be associated with many factors, including the type of stone as well as its location on a building and levels of exposure to different elements (Figure 7.12). The location also determines how rain or prevailing winds carrying particles reach the surface. The wetting and drying cycles are significant to the weathering of the surface, which will increase susceptibility to frost damage, while airborne pollutants can cause surface spalling. The process of deterioration and suitable conservation interventions required can be very different for different types of stones. Techniques used in the conservation of limestones and sandstones are often different.

Although stone has become a cladding material in most contemporary applications, for historic buildings it is likely to have been used for structural purposes as well. Structural failures in stone include the failure of elements such as arches, vaults and lintels (see Figure 6.12), or the delamination and buckling of walls. Different stones age in different ways. Microbial colonisation of porous stones occurs when conditions are right, subsequently damaging the weathering crust and exposing the softer stone. Moisture in combination with salts is a common cause of deterioration of sandstones.

Iron cramps, used since antiquity, are a method with which stones were connected to one another and stabilised in a structure. When in contact with water, iron can rust and expand, eventually cracking the stone surface. Rust staining on a stone surface is usually an indication of rusting fixings. If too close to the surface of a stone, the expansion of rusting metal can also cause cracking and spalling of the surface.

Stone paving may suffer from subsidence caused by an unsuitable base, or cracking and movement caused by heavy vehicles or tree roots. Stone paviours are often an integral part of the character of a historic town which is lost when replacement is undertaken with uniform concrete paviours.

Principles and techniques of repair

Structural repairs to masonry walls include:

- The use of metal dowels and rods or reinforced concrete to stitch a buckling wall (see Chapter 6)
- Grouting to fill and consolidate loose masonry, especially where there are rubble cores (see Chapter 6)
- Replacement or repair of individual stones where the decision will need to be made whether to maintain, replace, piece in a new section or carry out a mortar repair

Cutting out and replacing is the traditional method of stone repair, but with every removed stone, valuable historic and archaeological evidence, such as tool marks, masons marks and moulding details, will be lost. Any new stone being inserted into an existing stone structure should be from the same source or, if this is no longer possible, a close match to enable similar performance should be considered. Where it is necessary to differentiate between old and new stone, this is often achieved through the contrast between the new and weathered old stone or the surface and decorative treatment of the new stone (Figure 7.13). Replacement should be made where structural stability is a priority. With the escalating expense of access and scaffolding, site decisions tend to favour replacement. Another approach is to remove only part of the stone that has weathered most, cut in a replacement and fix it with a stainless steel dowel or cramp and resin. Piecing in can also be used to reface a stone, where the decayed surface is replaced but the core is retained in situ. Where mouldings are being copied for replacement, this should be based on the known original. For mortar repairs the mortar mix must be compatible with the stone (different for limestones and sandstones). Mortar repairs have a lifespan of approximately 30 years, which is considerably shorter than a stone replacement (Figure 7.14).

All great stone buildings have been subject to repair and renewal programmes over several centuries. The cutting of stone using traditional masons' tools from quarrying to fine carving is so labour intensive and costly

Figure 7.13   Stone conservation at the Temple of Trajan, Pergamon, Turkey. Missing sections are replicated in artificial stone using a mix containing marble chips close to the original stone. Mouldings are cut as a profile only and not in the detail of the original.

Figure 7.14   Cutting out mortar using power tools damages surrounding masonry.

Figure 7.15 New replacement stones contrast with the substantially weathered wall surface, but are 'honest' copies of the original.

that today much of the cutting will be done by machine. This creates a disparity between historic hand-cut stone and the sharp edges and precision of contemporary machine-cut stone. It is sometimes the case that the original quarry is not known or is no longer open. Stones that are not from the same source might appear to be a match at the start, but over time will weather differently from the original. Similarly, artificial stone weathers differently and will form a different patina from natural stone. When stones have been replaced, the mosaic of older and new stones rarely presents a uniform appearance, but the new stone must be allowed to weather at its own pace over time. New stones are also likely to stand proud of the existing weathered surfaces (Figure 7.15).

The use of thin clay tiles in the repair of stone masonry was pioneered by the Society for the Preservation of Ancient Buildings (SPAB) as an 'honest' method of repair. They are sometimes limewashed or rendered over. Although it may work as a technical solution, it remains debatable as to whether it is an honest repair or a detriment to the aesthetic unity of a building, especially after a substantial amount of repairs have been carried out.

New materials or products that are used with stone include protective coatings to improve water-repellent qualities, or substances such as epoxy resins used to consolidate, restore or replace damaged sections. The application of consolidants to prevent accelerated cases of decay on stone surfaces has not always been successful and has at times increased the rate of decay.

Figure 7.16   A stone roof being relaid on a modern breathing membrane in Stromness (Orkney), UK.

Some coatings sold as waterproofing actually block pores, and can be especially problematic when applications are not reversible. Products that have not been fully tested may also damage the original material in the long term. Only treatments with a proven track record provided by a reliable source and applied by a qualified contractor should be used. Where surface treatments are recommended, consideration also has to be made for how long they will last and how often a repeat treatment might be needed. For example, lime-casein shelter coats applied to limestone as a protective coating would need to be reapplied every ten years. This will have cost implications for maintenance budgets, especially for parts of a building that are difficult to access.

Traditional stone tiled roofs are increasingly being replaced with manufactured roof coverings. Damage to stone roof coverings can be as much the roof structure and base as the stone tiles themselves. If complete reroofing is necessary, then all stone tiles should be removed, the timber base made good and a layer of modern roofing felt applied to allow for the roof to breathe. The stone tiles can then be refixed, reusing all the good stones and replacing the remainder with new ones to match (Figure 7.16). Wooden pegs used to fix stone or handmade clay roof tiles are now likely to be substituted with stainless steel fixings. Slates are a more commonly used roofing material and have also been copied in artificial versions and even plastic roof vents. Artificial slates are uniform in appearance and their use in conservation projects is not recommended.

Figure 7.17   A traditional brick clamp of hand-made bricks after firing near Mecca, Saudi Arabia.

BRICKS AND CLAY ROOF TILES

Building in brick

Bricks are a product of clay and sand that are moulded together and usually fired. The colour, shape, dimensions and quality of bricks vary considerably through time, location, material source and manufacturing techniques (Figure 7.17). At the temperature at which bricks are fired, the material and any chemical constituents of the clay give bricks their colour. Bricks are laid in mortar to form walls as well as structural forms such as arches and vaults.

In areas where clay is abundant, brick is the common and cheap building material from which a large proportion of the building stock will be constructed. It was common in Roman times to build massive structures out of brick and then clad them in thin sheets of marble. Throughout history, brick structures either have been clad in stone or rendered over to give the impression of heavy masonry structures. Brick continues to be used as a versatile building material, especially in domestic buildings. In other instances brick has been used to create decorative surfaces, especially in the nineteenth century where gauged brickwork was elaborately moulded or combined with terracotta. Gauged brickwork is made from bricks that are set with very slim joints using a lime putty and can be cut to size and/or carved. Gauged brickwork requires highly skilled craftsmanship and is rarely used in modern building applications other than for lintels and arches.

Causes of decay and failure

The failure of brickwork can arise from faults in the building structure, faults in the bricks or through moisture penetration. It is always possible that bricks used for construction are faulty in that they contain impurities, are warped or have been underfired. This may be the case for an entire batch or only some of the bricks. Movement in the building structure can cause damage and cracking of brickwork. As long as brickwork is set in an appropriately weaker mortar, small amounts of movement will be taken by the joint where the cracking will appear. Where a crack runs through the brickwork, structural advice should be sought.

Water penetration has serious impacts on brickwork, as bricks will break down as they become saturated (Figure 7.18). The resulting crystallisation as salts move to the surface causes spalling. Bricks of a porous consistency are more likely to be affected by frost, whereby damp that has penetrated the brick will freeze and expand causing the surface to spall. As the joint mortar is lost, water sitting in the open joints will weather the brickwork. Where brickwork is laid in a hard mortar, it will wear away from the joint.

Principles and techniques of repair

Much of brickwork repair involves the cutting out and replacement of single bricks or sections of brickwork and repointing. Where structural damage has occurred, anchors or steel ties may need to be inserted into the brick wall and the cavities grouted. A facadist approach to modernising property in the eighteenth-century England was to build new brick facades onto timber frame buildings. The consequences of this practice have been delamination or buckling that has to be stitched back using steel ties.

One of the first considerations of brickwork repair is to obtain bricks that are of the same size, quality and colour as the original. In Britain, recent

Figure 7.18   In this example the string course of brick has been damaged by the failure or movement of the lead capping. Further damage has been caused by the insertion of a new downpipe and marks left from plant growth can be seen on the wall. Replacement and possibly new detailing of the lead protection and securing of joints to stop any further water penetration would maintain the string course, although some heavily damaged bricks may have to be replaced.

changes to metric measurements means there is not a ready supply of imperial-sized bricks available. The use of salvaged bricks has been an option, but with increased listing and conservation area designation, less and less original material is coming on the market and there will always be questions of provenance and authenticity. Salvaged bricks must be clean of any mortar and this is possible only if they have been bedded in a lime mortar, and the arrises will need to be in reasonable condition. For a large conservation project, it may be possible to specially manufacture bricks of the right size and colour and there are suppliers that will produce bricks using traditional methods, including hand moulding. Modern factory-made bricks are usually too uniform for the conservation of older buildings. Where the face of a brick has spalled, it is best to cut out the brick and replace it with a new brick. It may be possible to reuse the same brick placing the backside to the front.

Traditionally, brick walls would have been constructed using a lime mortar. The natural cycle of lime means that the mortar will eventually deteriorate and loosen out of the joint. It is quite natural that a brick wall will need to be repointed from time to time. When repointing, the original mortar mix, the style of pointing and the condition of the brickwork needs to be considered. Any evidence of the original pointing technique should be recorded. The most appropriate mortar for pointing is a lime and sand mortar, unless it is a high-exposure location or joints are vertical such as in paving, in which case a hydraulic lime mix may be more appropriate. Cement-based mortars

Figure 7.19   The pointing on this garden wall has been carefully executed to be flush with the now worn arrises of the brickwork.

are best avoided in most situations. The existing mortar will also need to be raked out to a depth that will ensure the new mortar will hold. Since some of the original sharp arrises will be lost or worn, the new mortar should be flush with the existing edges (Figure 7.19). Cutting out bricks for replacement or removing mortar for repointing can be time-consuming, but the use of power tools must be avoided, so as not to cause damage to surrounding brickwork (Figure 7.14). Replacement bricks must be fully tied into a wall. Where it is not possible to cut out a damaged brick, a thin facing strip called a slip or a mortar repair might be used.

The repair of gauged brickwork is complex and requires highly skilled craftsmen to undertake it. Patching or plastic repair is most likely used in this type of brickwork. As long as the colour match can be achieved, this is an appropriate and cost-effective approach that uses a mix of brick dust, lime putty and sand. Mortar repairs are also appropriate where a hole caused by an earlier fixing needs to be filled in a brick.

The repair and replacement of clay roof tiles will follow much the same principles as the replacement of slate roof tiles explained above. Traditional clay roof tiles should not be replaced with machine-made tiles and tiles of a different shape, dimension or colour (see Figure 7.25). Where tiles are fixed with mortars, then as with brickwork, an appropriate soft mortar should be used to enable movement (Figure 7.20).

Figure 7.20 The necessary replacement of roof tiles in Dubrovnik following heavy bombing impacts on the townscape character of the old town.

TILES, FAIENCE AND TERRACOTTA

Tiles, faience and terracotta on buildings

Terracotta and faience are also fired earth products. They are made from pure clays and moulded, creating decorative elements that would not be achievable with bricks. Terracotta became popular in the nineteenth century in England, often used in conjunction with brick. Terracotta is also seen on the early skyscrapers of Chicago, USA, where it was used as a cladding material on the steel frames. Terracotta could be fired as a hollow box which was then filled on site, once the sections had been connected to the substrate and one another using metal fixings.

Faience, fired as a solid block, is more akin to a tile, including fixing with a mortar bed. The most common application of faience is in interiors where they are often glazed. Although these two materials are commonly associated with construction in the latter half of the nineteenth century, terracotta is one of the earliest known materials and has been a means of producing decorative adornment for buildings since the early Greek temples. Very little terracotta or faience is produced today.

Ceramics are also fired clay products, where different raw materials, techniques and firing temperatures create different products. Ceramic tiles continue to be produced and have a range of internal and external applications from floors to kitchen and bathroom surfaces to cladding buildings. Examples

Figure 7.21 Sections of the recently restored encaustic tile floor of St George's Hall in Liverpool, England, has to be protected to avoid any further damage. (Photograph by Simon Woodward.)

of ceramic tiles used as flooring or wall coverings still survive from antiquity and medieval times. Different manufacturing methods and application of decorations affect both the level of deterioration as well as the conservation techniques that need to be applied. Encaustic tiles were made with inlays of different coloured clays, made popular again in the nineteenth century in England and were used for both floors and walls. They are, however, subject to wear, especially when used as floor tiles, and some outstanding examples now need to be protected with overlays (Figure 7.21).

Figure 7.22   Tiles falling off from one of the Central University of Caracas buildings, a conservation concern across the World Heritage Site Modern Movement campus in Venezuela. (Photograph by Eddy Butron.)

Causes of decay and failure

On firing, whether with a glaze or not, the clay acquires a fireskin which acts as a protective coat to the softer clay inside. While this skin is reasonably tough, once it is damaged the rate of decay is accelerated. Water ingress causes disintegration, especially in the softer clay backs, and can also rust any metal clamps or fixings. Rust and corrosion of fixings can shatter terracotta. Water ingress behind a glazed surface may cause it to spall and if the glaze is porous or defective as a result of the manufacturing process, then cracks may occur and the permeability of the material increases. Failure of mortar joints and the different movement patterns between terracotta and its substructure is another cause of cracking. Although ceramic tiles appear to be hard wearing, they are also liable to deterioration, crazing, chipping and cracking caused by structural movement and absence of movement joints, water and frost damage, efflorescence and harsh cleaning methods and substances.

Tiling using ceramic tiles or smaller mosaic tiles became a popular cladding method in the twentieth century. The failure, particularly of adhesives and the subsequent loss of tiles, has created a new conservation problem (Figure 7.22). A particularly tricky problem is when the original material is no longer man-ufactured and a suitable replacement in material, colour and dimension can no longer be sourced. Even a slight change in colour of a repair material produces a very different effect than that intended in the original design and is one of the challenges in their conservation. In some cases a new overcladding has been proposed, maintaining the original material underneath. While in conservation, this enables the preservation of original material; it will make a significant change to both the volume and the appearance of a building.

Principles and techniques of repair

The vulnerability of the fireskin means that even cleaning methods adopted for terracotta surfaces need to be carefully selected and trialled. Abrasive methods of cleaning or the use of acids are not recommended. On glazed surfaces soiling can be removed using water, but it is impossible to remove any soiling that is beneath the glaze without causing further damage.

Loose tiles and pieces of terracotta can be fixed back using stainless steel pins and epoxy resin, whilst small areas of damage can be filled in with mortar repairs. For this, either a masonry cement, sand and aggregate or a lime and stone dust mixture can be used. Colour sampling must be undertaken to ensure that the repair will match the original in colour. The colour of glazing can also be matched in repairs. Tiles should be restored in situ wherever possible. Broken tiles can be bonded using an appropriate adhesive and gaps filled using a filler. However, while this is appropriate for interiors, it may not be sufficiently durable for outside surfaces or for floor tiles. Detaching tiles for conservation purposes can cause further damage and should be avoided wherever possible. On decorated surfaces, cleaning with soft brushes is the most appropriate cleaning method.

Where entire blocks of terracotta have to be replaced, these will need to be specially manufactured. The replacement element will need to be of a clay that will colour match and be produced in a way that will allow the correct amount of shrinkage in firing. Replacement using other materials is not desirable or advisable as these materials will weather in a very different way to terracotta. There will often also be a visual disparity between the original, weathered terracotta and the new replacement infills. Nonetheless, colour matching needs to be on the basis of the original and not the weathered material. In order to avoid future cracking, all corroded fixings should be removed and new components attached using stainless steel fixings that are bedded in a mortar which is weaker than the terracotta.

STRUCTURAL METALS

Structural use of metals in buildings

The metals commonly used for structural purposes are iron and steel. Iron, found in the earth as an ore, is processed to produce the metal iron. Wrought iron is the most common form of iron used from antiquity through to the

Figure 7.23   An iron column and bracket support a timber beam in this Liverpool warehouse.

eighteenth century. It is the purest form of iron and is versatile in that it can be welded and worked in hot or cold forms. The Industrial Revolution brought a greater demand for iron and therefore for more efficient methods of production. This resulted in cast iron, which unlike wrought iron cannot be worked and is dependent on the form it takes in the cast. One of the first uses of cast iron was as supporting columns in warehouse construction, since it is strong in compression, unlike wrought iron which is high in tensile strength (Figure 7.23).

Steel has a higher carbon content than iron and therefore can be hardened in water once it has reached bright red heat. Alloying steel with other metals can increase its toughness and resistance to corrosion. Steel became a material of the twentieth century, used to produce the steel frames that made high-rise building possible and as a reinforcement in concrete.

Causes of decay and failure

Metals will start to corrode as a reaction with moisture, pollutants, concentration of acids and salts or other materials and metals found in their immediate environment. In ferrous metals, corrosion appears as iron oxide or rust. Rusting metal often gives the impression that most of the material has turned into flaking layers of rust, but oxidation increases the volume of the material and wrought iron can expand up to seven times its original volume. This is most problematic when metals are incorporated into a building's structure and start causing damage and movement in surrounding materials. Rusting reinforcement bars will cause concrete to crack and fall off. In high temperatures, such as in a fire, metals lose their strength and distort or fail.

In steel frame structures, water penetrating through the cladding is often the cause of corrosion. In most instances, as long as the cause of water penetration is remedied and the rust removed from the steel, it is unlikely to have weakened the steel to a level where additional strengthening is required.

Principles and techniques of repair

Methods of repairing cast and wrought iron are welding, fitting new straps, pins or dowels, replacement sections and the use of fillers. Options for structural repairs include strengthening by adding new members, including columns, but the aesthetic consequences of this and its impact on the spatial qualities and significance of the building will need to be carefully considered. New castings can be made in an iron foundry as long as an existing component can be used to generate the mould.

Most iron surfaces will have been treated in one form or another for rust and as long as these are sound, new applications can be applied over them. Any rust must, however, be carefully and fully removed prior to further coats of finish being applied. There are many methods of cleaning ironwork including manual, mechanical and chemical approaches. When undertaken with care the use of a flame to remove rust and loose scales is one of the more effective measures that can be carried out on site. Other options include the use of abrasive methods, but care needs to be taken where decorative elements might become damaged. Cleaned ironwork must immediately be treated with a preservative to avoid further rusting. An option for cast iron or wrought iron structures may be to dismantle, clean and repair the components before re-erecting them on site. Smaller items, such as gates or railings, can be removed from site and repaired in a workshop. Care needs to be taken when dismantling fixings, including where they are fitted into another structure, such as railings that are set in a coping stone or gate post.

Fire safety for exposed iron or steel components can be provided with the application of an intumescent coating that will foam up in the incidence of a fire to provide an incombustible layer around the structure.

SHEET METALS

Sheet metals in buildings

Nonferrous metals including lead, copper, zinc and aluminium are commonly used in sheet form as roof coverings, cladding and other weathering details. Lead is a soft material that can be easily shaped with the use of hand tools. Lead has been used in molten and sheet form as a building material since antiquity. Molten lead was poured into the joints of Greek and Roman temples to set the iron cramps and dowels (see Figure 7.11). Its most common use, however, is as a roofing material and for pipework (Figure 7.25). Most large domes, such as some of the great mosques, are covered in lead, while the original Roman bath in Bath is lined in lead. Lead cames hold together the panels of glass in stained glass windows and windows that are made up of smaller panes of glass.

Lead sheets were traditionally prepared by casting molten lead on a bed of sand, but since the nineteenth century most lead sheeting has been milled. Milled lead sheets come in different thicknesses that are used for different purposes, with thicker codes used for roof coverings while thinner ones are used for flashings and similar detailing work. Copper and zinc sheet are also used as roofing materials, but neither is as pliable as lead. Zinc, another hard metal, has also been used as a roofing material in the form of tiles. Bronze, an alloy of copper and tin, was more commonly used for fixtures and fittings in buildings and for statues.

Corrugated iron was developed in the nineteenth century as a convenient material for prefabricated buildings that could be shipped around the world. The metal sheet cladding and roofs were complemented with modest ornamental details also made from metal. Many churches, halls and other buildings conceived in the UK were shipped to colonial outposts including Australia and South Africa (Figure 7.24).

Causes of decay and failure

The major cause of failure of lead roofing or cladding is a result of poor fixing, the use of sections that are too large, too many fixings or the failure of fixings, all of which cause distortion or sagging of the material. Initial corrosion of lead leaves a thin grey patina on the surface that serves as a crust which will protect the material from further damage. However, condensation that occurs on the underside between the substructure and the lead is likely to cause corrosion of the inner side, including on newly laid lead. The life of a lead roof is from 50 to as long as 250 years.

Figure 7.24 A corrugated iron church, now housed at the Museum of East Anglian Life (Stowmarket, England).

Chemicals in rainwater or the polluted atmosphere will cause corrosion of copper. Most typically copper sheeting is damaged because of thermal movement, accidental damage or wind lift that causes the thinning and cracking of the material. Deterioration of the substructure is one of the common causes of failure to copper roofing or cladding.

Principles and techniques of repair

Repair techniques for the various sheet materials will differ, but generally the repair of sheet materials involves making good joints, opened up seams and welts or replacing missing clips and placing additional clips where necessary; carrying out patch repairs soldered on to the metal and adding new sections of material, such as between two bays.

Patch repairs might prove a short-term solution or a means for repairing a small hole. Lead burning can prove to be a useful technique for repairing and fixing lead in situ, but the use of open flames on works concerning historic buildings may not be permitted, especially since several major fires to historic properties in recent years have been caused by torches used by roof workers.

While copper and lead seams can be unfolded relatively easily to remove damaged sheets and replace them with new ones, this is more difficult with the harder zinc standing seam. Present-day building advice involves the use of

Figure 7.25   Lead is used as a roof and cladding material as well as for soakers and flashings on a tiled roof.

smaller sections of lead and the repair or replacement of a lead roof is likely to take on some variation to the original, including incorporating lightning protection. Lead can also be used as a repair material, especially as capping on vulnerable masonry. While a replacement lead sheet will blend in with existing sheets in a reasonably short time, the same is not the case for copper, where it can take some time for the new sheets to weather to a similar colour as the original sections. Modern techniques do provide a solution in the form of pre-patinated copper sheeting. On lead the use of patination oil on completion will avoid surface staining (Figure 7.25).

Corrugated iron buildings are mostly modest buildings and therefore less likely to be protected; this also means that a great number have now been lost. Repair techniques usually involve the replacement of individual panels. However, original panels will have a history of paint and dents which are also part of the history of the building, and replacements should only be used where members are severely damaged or rusted.

Bronze statues are best treated in a metal conservator's workshop where any corrosive material can be removed from the surface, cleaned, damaged sections repaired and the bronze rewaxed. The use of resin coatings on bronze has often been the cause of further damage and the removal of the material is not possible without damaging the bronze surface.

GLASS

Glass is a man-made material. Because of its brittle and fragile nature and the technologies involved in its production, glass as a building material has gone from being used in very small panes to increasingly larger glazed window surfaces, to the substantial sections of unframed and structural glass that is used on new buildings today. Glass has thus played an important part in informing building design through the ages and the treatment of facades.

The conservation of glass ranges from the delicate and highly specialised concern of the repair and maintenance of stained glass to the challenge of finding suitable replacements for glass products, such as Vitrolite, produced in the 1960s but no longer available today.

When conserving historic windows, the basic principle is to retain as much of the original glass as possible as this retains much evidence of earlier production methods, and characteristics of lucency, light reflection, colour and texture that would not be replicated in a modern-day glass. Even the fitting of external protections over fragile windows can significantly change the perception of a window internally and externally. Another challenge, however, is achieving present-day energy performance targets. Where a replacement has to be made, it is possible to obtain sash windows (and metal windows) in the same profile but incorporating double glazing. In Sweden, the renovation of existing timber windows, rather than their replacement with modern triple-glazed variants, achieves competitive U-values by using a new glass with low-emission coating. The conservation of stained glass is a specialist area and should be carried out in a specialist glaziers' workshop.

RENDERS, PLASTERS AND MORTARS

Renders, plasters and mortars in building

Mortars are the softer materials that are used to bind masonry components and in certain instances used as a render on walls and ceilings for protective and decorative purposes. Earth-based mortars have been discussed in the earlier section on earth structures (Figure 7.26). Most mortars are made up of binders such as lime, gypsum or artificial cements that are then combined with sand and aggregates. To add strength to mortars or to increase their cohesive properties, traditional mortars have often included additives such as animal blood, hair, urine, egg white or date pulp. The mortars used for jointing must always be the weakest element in a structure so as to allow for some movement and settlement without cracking the masonry. Mortars that are harder than the masonry they bond will increase the rate of decay as the stone or brick will wear away from the harder joint (Figure 7.4). The harder

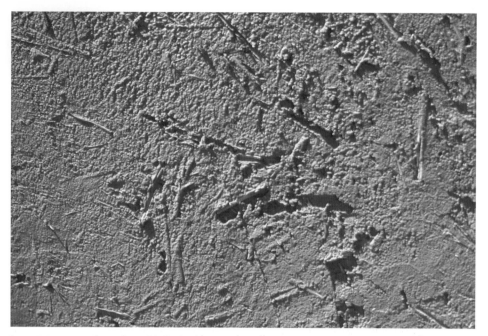

Figure 7.26   Straw is used as a binder in this earth mortar in Mexico. (Photograph by Nuray Özaslan.)

the material or stone, the harder the mortar can be, so long as it remains softer than the masonry.

Lime, a key component of most traditional mortars, is produced from burning limestones, sea shells, coral or marble. The burning of the material in a kiln produces quicklime, which is then slaked with water to produce lime putty. Slaking lime is a potentially hazardous process that requires due care and is one of the reasons why the use of lime has become less common. Lime for mortars can also be obtained in a dry powder form of hydrated lime.

The uses of additives such as ash or brick dust will increase the hydraulicity of limes. Hydraulicity refers to the ability of the mortar to set under water and is most important for engineering purposes. The most common use was the addition of volcanic ash, pozzalana, to lime to give it a hydraulic set. Different stones produce limes with different degrees of hydraulicity. Limestones that contain clay will burn to produce hydraulic lime. Hydraulic limes come as a ground hydrate and, unlike nonhydraulic limes, once mixed up into a mortar must be used immediately. The properties of a mortar prepared with lime putty and with hydraulic lime are very different.

On internal plasters, decorative features were also created using gypsum, also known as plaster of Paris, and paper-based products such as *papier mache* or *carton piere*. Gypsum is also used for the creation of decorative screens or ornamentation of external surfaces, especially on earth structures.

Causes of decay and failure

Mortars, being the softest material in a structure, are also the most likely to be damaged and in need of repair or replacement. The lime cycle results in the breaking down of lime and all mortars with a high lime base will require replacement at some point.

Renders and plasters are most likely to get damaged in parts of a building where there is heavy traffic, such as in entrance ways. The render or plaster may start crumbling, cracking (often caused by shrinkage) or crazing. It may also lose its bond to the backing and become separated. This is most likely to be a masonry surface for external renders and timber laths on interior walls and ceilings.

Principles and techniques of repair

A repair mortar, render or pointing mix will need to match the properties of the original as well as its colour and texture. Finding the exact mix of a historic mortar requires laboratory testing and even then it might not be possible to establish an exact mix. Most of the material is likely to have been sourced locally. Secondly, the new mortar needs to be successfully keyed into the remaining original mortar as well as the backing.

Plasters and renders will have been built up in layers and repair work should follow the same principles. Ideally, repairs should be carried out as patches, where the defective material has been cut out to a regular rectangular shape. Partial and patch repairs are possible only if the cause of damage is not separation from the backing. For ceilings, where there is access from the floor above, plasterwork can be fixed back using a new mesh fixed from the floor joists. This is a tricky procedure and care needs to be taken to ensure the plaster is evenly supported from below.

For repointing, existing mortar should be removed only if it is damaged and causing deterioration of the masonry or if is unsuitable (e.g. hard) and causing damage to the masonry. New mortar should match the original in mix and aggregate content and suit the condition of the masonry, and all aggregates should be properly graded and washed. For pointing, loose mortar must be raked out of the joint to depth that will provide sufficient key for the new mortar. The pointing should go up as far as the edge of the joint but not over the masonry (see Figure 7.19).

The use of cement for repairs, grout, mortars or plastering is not desirable since it sets too hard and is not compatible with weaker and softer traditional mortars as it fails to move with the building, preventing the fabric from breathing. A hard mortar is difficult to remove and removal can cause more damage to surrounding masonry. Several other contemporary products are, however, used in conservation projects. Polyvinyl acetate (PVA) is a bonding agent used to help a repair mortar key into or adhere to the backing. High temperature insulation (HTI) is a white refractory brick powder that acts like a pozzalano and assists in the setting of mortars.

CONCRETE

Building in concrete

Early forms of concrete were known to the Romans who mixed pozzalano, a volcanic ash, with sand and lime into a mixture that could be moulded. The dome of the Pantheon in Rome is probably the most impressive use of the material at that time (see Figure 3.1). Although concrete as we know it today was first developed in the nineteenth century, it became one of the defining building materials of the twentieth century. Concrete depends on steel reinforcement to withstand tensile forces.

Causes of decay and failure

Research into the deterioration and repair of many twentieth-century materials is still in its infancy. In reinforced concrete, concrete deterioration caused by sulphate attack, freeze and thaw cycles, etc., and steel corrosion that occurs either when protection provided by concrete is lost or something is present in the concrete mix, such as salt-contaminated aggregates, are the major conservation issues that need to be addressed. The concrete may have been of a poor quality to start with, may not have been compacted sufficiently or the reinforcement bars placed too close to the surface. Design faults like too wide spans between expansion joints can cause cracking (see Figure 6.2).

Flat roofs are another common feature of the modern movement and probably the one most likely to fail. Where the roof construction is determined by a single waterproof membrane, major areas of failure have been the short lifespan of some of the materials used and weaknesses at joints, especially at overlaps or at the upstands.

Principles and techniques of repair

Apart from major structural repairs, most localised repairs to reinforced concrete include crack injection, surface treatment or concrete replacement. Where reinforcements have rusted and are causing the concrete to crack or spall, the concrete needs to be cut back and the reinforcement bars carefully cleaned, often by grit blasting. Once cleaned, the bars will need to be treated with an anticorrosive layer and the concrete recast using localised formwork. Most important will be the removal of the cause of corrosion or deterioration before undertaking repair works.

One way of reducing the threat of corrosion of the internal reinforcements in concrete is to use cathodic protection. This method has been developed to protect steel against corrosion and can also be used for reinforcements within concrete. The system is based on the transmission of a continuous electric current through the structure that ensures the protected metal components do not reach the electric potential that will enable the process of corrosion. Cathodic protection depends on a power supply to provide a current and

Figure 7.27 The La Tourette monastery in France, designed by Le Corbusier, where patch repairs on the barefaced concrete are too obvious and detract from the integrity of the surface.

therefore requires regular maintenance and monitoring. Anodes will also need to be placed in locations where they do not detract from the architectural qualities of the building, but provide sufficient protection and continuity between all steelwork components.

Matching the colour of new concrete to the existing is often problematic in barefaced concrete, particularly if it is board-marked, i.e. holding the imprint of the timber formwork (mould) on its surface (Figure 7.27). Nor is patina, celebrated on older buildings, so welcome on concrete surfaces, possibly because the materials does not age well or the building style is too familiar and the aging is equated with a shabbiness that would not be attributed to an 'aged' earlier period building. Non-permeable coatings also need to be avoided on concrete surfaces as they will trap moisture inside. Even in twentieth-century buildings, the full-scale replacement solutions, recommended for some failures, need to be rethought since smaller scale and more appropriate approaches coming out of a detailed analysis and understanding of the building will be environmentally less costly.

PLASTICS AND RESINS

Plastics and resins are another more recent building material that are now also the concern of building conservation. Plastics and synthetic resins are products derived from petroleum and include recognised building products such as asphalt, nylon, vinyl, polyester, silicone and Bakelite. Some plastics were developed to replace the use of ivory and bone in use for everyday objects, including some building components. Resins themselves have been used as conservation materials for some time. However, since their characteristics can change over time and under different conditions, they cannot be seen as a long-term solution. Yet, resins are also difficult to remove and repairs undertaken with resin are rarely reversible.

Not all modern materials have proven to be the runaway success they were claimed to be earlier. For example, materials like asbestos and particle board, widely used in a whole manner of applications such as insulation and fire protection, are now known to be a major health hazard. The safe removal of asbestos from a building has become a costly preliminary outlay in many conversion and rehabilitation projects.

FINISHES

Materials used for coatings and finishes

Many building materials were covered internally and externally in a number of different kinds of coatings, both to protect them from the elements as well as to provide decorative finishes. In many places around the world, timber was protected with oils or paints containing natural minerals that would also act as pesticides. A simple limewash, for example, will also deter insects. On the other hand, it became fashionable in the Renaissance to paint stucco on elevations to appear like stonework (Figure 7.28). Since the outer coat is the most vulnerable, it is also the layer of a building that will have been most regularly replaced. A decorative scheme is also the cheapest way of updating an interior or even an exterior to keep up with new fashions. Excavations of Roman villas have revealed several layers of mosaics laid over one another that may have been undertaken either to remedy a damaged floor or simply to reflect changing tastes.

Internal decorative finishes range from timber panelling and tiles, as discussed above, a variety of paint and plaster finishes such as scagliola, a plaster that is coloured and mixed with pieces of coloured marble and polished to imitate marble, frescoes or trompe l'oeil, to other materials such as textiles, silk hangings, leather or paper wall finishes and carpets. Various paper-printing techniques go back to the sixteenth century and became popularised in Britain during the Arts and Crafts movement of the late nineteenth and early twentieth century. Fabrics were often held on frames. Plain paint finishes involved the use of distempers that used animal glue as a binder and lead-based oil paints.

Colour schemes throughout history were determined by fashion and the availability and cost of pigments. Some of these may not appeal to present-day tastes, especially where the historic significance is not understood. The

Figure 7.28   Renaissance façade in Slavonice, Czech Republic.

choice of a bright yellow in the eighteenth-century England was an indication of wealth rather than sophisticated taste, for yellow pigments were expensive.

Principles and techniques of conservation

The layering and build-up of surface coats provides valuable evidence on the history and evolution of a building and should be retained wherever possible and appropriate. However, where there has been a substantial build-up of paint (modern paints are much thicker than historic ones) that is obliterating a key design feature, such as a decorative moulding, then some degree of removal, following records being made, might be more appropriate. In other instances, an inappropriate layer may be contributing to the decay, such as reducing the breathability of the building fabric, and may need to be removed. Many modern paints sold for their weather-protective qualities often have an adverse effect on traditional buildings that depended on the breathability of the fabric (Figure 7.29). Otherwise paint layers are part of a building's archaeology and should be maintained as such, rather than being stripped each time the building is redecorated.

Historic buildings continue to be painted or decorated as part of regular maintenance cycles and in response to user needs. Where plain painted surfaces can be painted over, different approaches will need to be sought for more decorative schemes involving stencilling or frescoes (Figure 7.30). In some cases in conservation, the integrity of the interior depends on a historic

Figure 7.29 Modern plastic-based paints are now used to apply traditional designs to the walls of these mud brick houses in southern Saudi Arabia, inadvertently impacting on the properties of the building fabric.

Figure 7.30 The conservation of frescoes is a highly skilled specialism of art conservation. The conservation of these frescoes at a monastery in Romania is being undertaken by a team of conservators who have undergone a minimum of four years' training in material sciences as well as techniques of conservation and the art of making frescoes.

colour or decorative scheme. The recreation of an exact replica of a historic decorative scheme is likely to be appropriate only in a museum environment.

While the use of original materials will give a more authentic appearance and work better with the building fabric, this may not always be possible for practical reasons. For example, the use of lead-based paints, which are toxic, is discouraged, yet the slightly dull or matt appearance of lead-based paint cannot be replicated with modern paints. The use of traditional paints and methods of application will result in slightly uneven surfaces that modern paints have all but obliterated. In many instances a 'historic' paint scheme may not appeal to modern-day tastes, especially in interiors. Much of the paint finishes used on historic buildings today reflect present-day tastes rather than a concern for historic authenticity, and this is part of the adaptive reuse process. Even hundred years ago, none of the magnificently glossy black painted iron railings, lamp posts or balconies would have been black, since the amount of pigment required for the lead paint would have caused problems in drying. The picture on the front cover of this book shows the blue-coloured ironwork railings of the eighteenth-century house in Spitalfields in contrast to the black gloss of the 'heritage' street lamp. Similarly, a pure white is a colour that has been achieved only in recent times and most traditional windows frames would have been painted in darker colours like brown or green.

As in any conservation work, it is important that detailed research for historic finishes is carried out prior to undertaking works. Information comes from a variety of sources, starting from documentary evidence relating to the property, including paintings of it and contemporary written records. For a more detailed analysis, a cross-section sample that is embedded in resin ground down to expose the layers and then analysed with the aid of an optical microscope is used. The analyses of samples can happen only against a good knowledge and understanding of the use of paints and surface decoration in the given region at the given time in history.

Fabrics or papers have often faded by being exposed to light or damaged by moisture or insect attack. The backing frames may be damaged or warped as well as the substrates. Papers may become detached and peel away from the surface. The techniques of repair are linked to those of paper or textile conservation, and will sometimes need to be undertaken in a conservation studio or laboratory. In situ repair techniques include the use of Japanese tissue paper and starch paste. (See also case study box on Uppark in Chapter 3.)

Case study: Danson House, England

Gold leaf was used for gilding, both externally and internally, a sign of wealth that many 'shimmering' interiors attested to. Gold leaf maintains its 'sparkle' for a long time, an effect that cannot be repeated by the use of modern paints that end up as a dull metallic finish. However, there is a noticeable difference between areas of newly applied gold leaf and older material, especially if it has been painted over in time. Danson House, an eighteenth-century Palladian Villa in Bexley, England, was in a poor state when conservation work commenced in the late 1990s following years of neglect. Although the building had been substantially redecorated in the nineteenth century, much of the eighteenth-century work remained underneath and therefore the presentation of the house in an original eighteenth-century scheme in conjunction with its distinctive architectural style was seen as the most appropriate. In the saloon, considerable amount of gold leaf survived from the eighteenth century on the door surrounds and cornices. This was maintained and only missing sections gilded, providing an interior that may not be fully uniform in appearance but one that remained truthful to the evidence.

CLEANING SURFACES

There are two reasons for cleaning building surfaces, one is on aesthetic grounds including maintaining the integrity or original design intention of a building and the second is to remove the cause of surface dirt or organic growth that may be causing the deterioration of the surface. Buildings have a history and part of the history is the soiling that is an accumulation of several centuries of existence. The objective of cleaning should never be to return a building to an 'as good as new' state. As buildings weather and attract various airborne particles, a surface patina is formed that is deeper than surface soiling. A poor choice of cleaning method and overzealous cleaning can result in the loss of what has become a top protective layer, leaving the surface more vulnerable to soiling and deterioration, as well as the loss of important historic information.

It is important to understand the type of soiling and what it is likely to respond to before commencing cleaning. This should be followed by carrying out trial areas to test the application. Cleaning may very well include a mixture of techniques to deal with different types of soiling and intensity on the surface. The most common methods of cleaning involve:

- Water
- Pressurised water
- Steam
- Particle jet
- Abrasive particles
- Chemical cleaners
- Poultices
- Laser

The simplest form of cleaning is to use water that will soften water-soluble dirt. Water is most effectively applied to the building as a nebulous spray (JOS system), reducing the risk of water penetration into the building and avoiding large quantities of water run-off. This technique works best on hard brick, limestone, marble and granite surfaces. Where surfaces have delicate paintings on them or tiles that are vulnerable to washing, dry methods are advised, and this work should be undertaken by a skilled conservator. The use of hand-held brushes or implements might also be needed for the cleaning of hard-to-reach details of mouldings and sculptures.

Abrasive methods such as sandblasting brickwork, a common approach in developer-led industrial regeneration projects, can leave bricks with an exposed porous surface that is more likely to gather dirt and deteriorate. The use of sealants to seal walls will also stop them from breathing and cause moisture to become trapped behind the sealant.

Chemical cleaners are either alkalis or acids. A weak hydrofluoric acid solution can be an effective way of cleaning some brickwork. The surface is pre-wetted, while surrounding surfaces are protected before the solution is applied, left on for a given time then carefully washed off. It is important that any chemical solution, whether it is for cleaning or a biocide against plant growth, is completely removed from a surface and that the run-off does not reach other surrounding materials.

Another method is the use of clay poultices that can also assist in the removal of stains or paint from surfaces. With this method a clay poultice containing minerals that will draw out certain staining is placed on the surface and as the clay dries out the dirt is absorbed into it. Depending on the level of soiling, several applications may be required. Recently, a technique combining a poultice with latex was pioneered in the cleaning of the internal surfaces of St Paul's Cathedral in London. The latex is sprayed onto the surface, and then peeled off once a sufficient dwell time has been achieved. This is followed with a more localised hand cleaning of details.

Graffiti is a concern on historic surfaces and often necessitates regular removal from the same surface. While the paint is usually easy to remove from most surfaces, some pigments can remain in porous materials. Rapid removal of graffiti is recommended and most of the substances used are more difficult to remove as they dry out. The use of any anti-graffiti coating, however, should not be damaging to the material.

Laser cleaning involves costly equipment and at the present time remains time intensive, and is therefore mainly used on smaller museum objects and sculptures. Laser beams may cause the discolouration of pigments on a surface.

SUMMARY AND CONCLUSION

Although different materials have been covered separately here, in practice they cannot be fully separated from one another as most building technologies and components are composites of several materials. The theory and principles discussed in Chapter 3 will apply to the conservation of materials, but with variations that need to be adapted to the different type of materials, their qualities and performance. The following are some general issues that have to be considered when making decisions for the conservation of materials:

- Understand that the building is a whole and works as such in structural and material repairs
- Consider the structural stability of the building
- Understanding the physical and chemical properties and characteristics of materials will inform repair techniques and procedures
- Where materials are concerned, information gained from records and observation may not be sufficient. At this point the use of a conservation laboratory to identify and test materials for strength and compatibility is strongly recommended
- The use of traditional methods should generally be the starting point for repairs
- New materials will only be acceptable where they are compatible, appropriate and have been adequately tested
- Seek a balance between maintaining as much original fabric as possible and addressing future maintenance needs (especially where access, opening up and scaffold costs need to be taken into consideration)
- Ensure conservation work is being undertaken by specialist conservators and that they are part of the team from the start of the project, since they can advise on how the conservation work will be undertaken and how it will impact on decisions regarding access, substrates and other materials, as well as the planning and sequencing of the project
- Consider health and safety issues and present-day building regulations at all stages

FURTHER READING AND SOURCES OF INFORMATION

Ashurst, J. and Ashurst, N. (1990) *Practical Building Conservation, Volume 1: Stone Masonry*. Aldershot, Gower Technical Press.

Ashurst, J. and Ashurst, N. (1988) *Practical Building Conservation, Volume 3: Mortars, Plasters and Renders*. Aldershot, Gower Technical Press.

Ashurst, J. and Ashurst, N. (1988) *Practical Building Conservation, Volume 5: Wood, Glass & Resin*. Aldershot, Gower Technical Press.

Ashurst, J. and Ashurst, N. (1991) *Practical Building Conservation, Volume 4: Metals*. Aldershot, Gower Technical Press.

Ashurst, J. and Ashurst, N. (1995) *Practical Building Conservation, Volume 2: Brick, Terracotta and Earth*. Aldershot, Gower Technical Press.

Ashurst, J. and Dimes, F.G. (eds) (1998) *Conservation of Building and Decorative Stone*, paperback edition. Oxford, Butterworth-Heinemann.

Baer, N.S. and Snethlage, R. (eds) (1997) *Saving Our Architectural Heritage: The Conservation of Historic Stone Structures*. Chichester, John Wiley & Sons.

Brunskill, R.W. (1992) *Traditional Buildings of Britain: An Introduction to Vernacular Architecture*. London, Victor Gollancz Ltd.

Caroe, A. and Caroe, M (2001) *Stonework: Maintenance and Surface Repair*, 2nd edn. London, Church House.

Charles, F.W.B. with Charles, M. (1984) *Conservation of Timber Buildings*. Cheltenham, Stanley Thomas Ltd.

Cox, J. and Letts, J. (2000) *Thatch: Thatching in England 1940–1994. English Heritage Research Transactions*, Volume 6. London, James and James.

Feilden, Sir B. M. (2003) *Conservation of Historic Buildings*, 3rd edn. Oxford, Architectural Press.

Lynch, G.C.J. (1990) *Gauged Brickwork: A Technical Handbook*. Aldershot, Gower Technical.

Moir, J. and Letts, J. (1999) *Thatch: Thatching in England 1790–1940. English Heritage Research Transactions*, Volume 5. London, James and James.

Oliver, P. (ed.) (1997) *Encyclopedia of Vernacular Architecture of the World, Volume 1: Theories and Principles*. Cambridge, Cambridge University Press.

Warren, J. (1998) *Conservation of Brick*. Oxford, Butterworth-Heinemann.

Warren, J. (1999) *Conservation of Earth Structures*. Oxford, Butterworth-Heinemann.

Watt, D.S. (1999) *Building Pathology: Principles and Practice*. Oxford, Blackwell Science.

Weaver, M.E. (1997) *Conserving Buildings: A Manual of Techniques and Materials*. New York, John Wiley & Sons.

English Heritage Research Transactions

Getty Conservation Newsletter

Journal of Architectural Conservation, Donhead

Web-based sources

Building Research Establishment (BRE): www.bre.co.uk

Copper Development Association, UK: www.cda.org.uk

Lead Development Association International: www.ldaint.org

The Lime Centre: www.thelimecentre.co.uk

Society for the Preservation of Ancient Buildings: www.spab.org.uk

Zinc Information Centre: www.zincinfocentre.org

Chapter 8

Regeneration, reuse and design in the historic environment

The concern of architectural conservation is not limited to the conservation of buildings but needs to be considered within the context of the urban and natural environment in which buildings are located and to which they contribute. Not all historic buildings may be exquisite architectural relics, but as a group they define and signify the character of a place and collectively they make up the historic environment.

At the same time, the historic environment is not a static entity but one that is continually changing and evolving so that it remains relevant and meaningful to contemporary society. Historic buildings are surprisingly versatile and have been adapted and reused over time, for a multitude of different and imaginative uses. At times, it is not just buildings but entire urban areas or groups of buildings that are rehabilitated and rejuvenated. The role of cultural heritage is often overlooked in urban regeneration, yet the historic environment can be an important catalyst in making regeneration successful, by adding the all-important ingredients of character and distinctiveness. Design plays an important role in the adaptation of historic buildings and places as dynamic living environments. This chapter considers the contribution of cultural heritage to regeneration based on an assessment of the economic value of cultural heritage. The second half of the chapter evaluates the various approaches to design in the historic environment, whether it is the extension of an existing building or an infill plot in a historic townscape.

URBAN REGENERATION

Conservation-led regeneration

Regeneration, like conservation, is a process, albeit one that can take much longer to demonstrate or to deliver change. Regeneration combines building reuse, urban design and new build projects within the framework of an economic development project. The regeneration of a canal or waterway, for example, will include nature conservation and biodiversity, regeneration

Figure 8.1   This project by Urban Splash in Liverpool's Rope Walks area combines the reuse of an old tea factory with the creation of a new public space.

and the local economy and transportation considerations. It often takes one building to be conserved and rehabilitated to create a trigger for others in the surrounding area to follow. As more buildings are regenerated and new users take up residence, the area will start to become regenerated and economic development take off (Figure 8.1).

For instance, Little Germany in Bradford, England, was transformed from being a largely derelict area at the start of the 1980s to a vibrant hub of cultural and business activity by the mid-1990s. The quarter, established in the latter part of the nineteenth century and originally named after the German immigrants who helped establish it as an important trading centre in the region, is characterised by grand warehouses and merchant quarters. The regeneration of the area involved public and private sector participation with the local authority initially focused on improving the public spaces, starting with the centrally located Festival Square, which also became a focus for a festival that helped place Little Germany on the regional map. A second initiative provided managed workspaces to encourage small business to move into the area. Today, the area has become a popular visitor destination and commercial centre, attracting over 100 businesses, while the historic warehouses are successfully being converted for uses including housing, hotels and offices.

In some cases, former industrial ports have been transformed into successful leisure destinations, often re-establishing a link between a city centre and its sea or riverfront (Figure 8.2). Many of these projects have taken different approaches to incorporating the historic port structures and warehouses into

Figure 8.2   Old warehouses now housing shops and restaurants add a distinct character to the waterfront in Sydney, Australia.

the regeneration schemes. Where the character of Baltimore in Washington, DC, one of the pioneer waterfront regeneration projects, is defined by the character of the older structures, in Bilbao, Spain, a new art gallery has been the focus of the riverside regeneration, while the reuse of historic buildings has been less visible. In Boston, USA, the conversion of Faneuil Hall, an old market building into an attractive contemporary marketplace became a catalyst in bringing a new leisure and entertainment focus to the downtown and harbour area.

Urban regeneration and urban conservation are closely linked. Whereas the objective or starting point of urban regeneration is economic and environmental development to improve physical and socio-economic conditions and the move towards a more vibrant and active environment, urban conservation is concerned with the conservation and rehabilitation of historic towns and areas, and remains more focused in the consideration of maintaining the historic fabric and character of a place.

Urban conservation

'The conservation of historic towns and urban areas is understood to mean those steps necessary for the protection, conservation and restoration of such town areas, as well as their development and harmonious adaptation to contemporary life' (International Charter for the Protection of Historic Towns, 1987).

Figure 8.3   The Royal Crescent in Bath, England, conceived as a uniform elevation, now has a distinct but different character as each house has been altered, cleaned and renewed in different ways over-time.

Urban conservation involves cultural continuity and the gradual adaptation of the urban environment. It is a political, economic and social concern: indeed more than anything, it is about people. A historic town encompasses tangible and intangible components within its built and social fabric. Up until the second half of the twentieth century, the major thrust of the conservation movement was monuments, architectural masterpieces or works of art (see Chapter 2). Many of the principles for value judgements and conservation principles predominantly related to sites and monuments rather than everyday buildings and the historic environments they are a part of. Although each individual building or structure might not be an outstanding example of architectural value, it contributes to the value of the group as a whole (Figure 8.3).

Writing in the nineteenth century, John Ruskin pointed out the value of historic cities, and the combined value of buildings, streets and spaces that made up the character of old towns, which he feared were being lost to modern developments. The example he focused on was Venice, where he proclaimed the two- to three-storey buildings on some of the smallest canals to be some of the best architecture. Yet, Ruskin's view was that of an outsider; one who was not faced with the burden of rehabilitating the damp buildings to which many of the inhabitants were confined to through poverty. This can be one of the fundamental conflicts of urban conservation where the values attached to an area by outsiders and the people who inhabit it can differ significantly.

Tourism today has become one of the prime reasons for the conservation of historic buildings and entire historic quarters. The reason is predominantly economic, but means that the view of the outsider can often be the guiding principle in conservation. Tourism often presents a greater incentive to improve the external appearance of buildings in a historic town, since most visitors will experience the historic character in the public realm. For most visitors, the cultural experience is not limited to historic buildings and monuments, but also includes traditions, a way of life, handicrafts, food and other less tangible aspects.

Alongside tourism, historic centres have become competitive commercial destinations, often acting as a regional retail centre. The amount, type and nature of the retail offer can impact on the character of a historic town. Many of the chain stores that now fill historic town centres provide globally recognised products rather than locally distinctive ones. Excessive signage can also detract rather than enhance local character.

The way in which a historic town or city centre develops is linked to the land use policies that are in place and that can be regulated by the local planning authority. While such plans are able to determine the mix of commercial and residential development for example, they cannot determine the nature of the shops (Figure 8.4). Often authorities are up against the economic pressures that allow commercial functions to expand. Other areas of concern are how the city centre is being used by the local population. In England, initiatives to bring derelict upper floors or properties in rear courtyards back into use have helped revitalise city centres. Many of these schemes depend on finding creative architectural and financial solutions to long-standing problems, such as the difficulty of access to upper floors, reluctance of shopkeepers to encourage mixed-use and city centre parking problems.

Traffic and circulation can be another restricting factor for historic urban areas that were not intended for motorised vehicles. Traffic and transportation impact on the historic environment in a number of ways. Most directly, vehicles driving down narrow streets or taking in tight corners regularly damage the historic fabric; vibrations can cause further damage while pollution from fumes is a known cause of soiling and subsequent material decay (Figure 8.5). More notable are the small changes brought about by paving front gardens to create parking space for cars that have an immediate impact on the townscape character of a street, reducing green space and contributing to more surface run-off water and higher localised flood risk. Yet, it may also be the only way of accommodating new uses and providing sufficient car parking spaces in dense urban environments.

In the improvement of the external realm in historic towns, there is often a tendency towards using 'heritage' style elements, many of which are standard catalogue products, whether they are 'Victorian style' lampposts

(a)

(b)

Figure 8.4   (a) A local café has recently been replaced by Starbucks at the gate to Canterbury Cathedral. (b) Globalised brands erode local distinctiveness in historic towns.

Figure 8.5 Parking and traffic flow can be a major concern in the narrow streets of many historic towns.

or signage. The choice of paving in the historic environment also needs to be carefully considered. It is often the case that the historically authentic surfaces are no longer suitable for use by motorised vehicles and sometimes heavy maintenance equipment. Where alternatives are sought, paving especially should not be of a colour or laid in a design that will detract from the historic characteristics of the streetscape. On the other hand, as noted in Chapter 6, surfaces increasingly need to cater for elderly and disabled users, in which case contrasting surfaces or railings may not necessarily fit in with the sought after character.

Urban conservation is primarily a process that seeks economies that will facilitate the conservation and sustainability of the historic environment. Where the revitalisation of a historic area enables the economic development of the area, funds still need to be secured for the conservation of individual buildings and in some cases the establishment of new uses that will give them a viable future.

Economic viability

With the exception of a small number of historic buildings of national significance, the conservation of most buildings is dependent on them having a valid and viable use or usefulness. For building owners, investment in maintaining a historic property is most likely to be undertaken when there is a perceived economic benefit, in the form of higher rental income or increase in property value. Hundreds of historic quarters around the world have been replaced by higher rise neighbourhoods, because the exchange of a dilapidated two-storey house for 'modern' apartments or commercial development was seen as a more attractive alternative for the owners. Changes in family structures from extended families to smaller family units have made many larger old houses redundant, at the same time creating demand for new types of accommodation. In some South East Asian countries, rent control acts have resulted in many protected shop-houses to fall into disrepair as the small rents received by their owners do not justify any investment in the properties. The perceived economic advantage of new build is also one of the reasons behind the numerous 'accidental' fires that consume historic buildings.

Although cultural heritage is an asset for a place and for society as a whole, its economic benefits may not be directly felt in terms of revenue generation. Heritage, like nature, also has a non-market value, and its benefits are more likely to be felt indirectly. One of these relates to the improved quality of life the historic environment can provide through its scale, character and a distinguished identity when compared to the uniformity and sameness of some contemporary living and working environments. However, while the value of a property may increase with listing, there will be additional costs that are incurred as a result of the conditions or consequences of listing. Listing will also limit the type of new functions a building can be used for.

At national level where budgets are limited, as they often are, decisions will need to be made on the amount and prioritisation of resources allocated to cultural heritage, as opposed to education or health for example. Buildings of a certain type or period to which higher value is attached by society may be prioritised over others, such as those linked to a colonial legacy. Tourism is often a reason for prioritising some sites or projects over others in an expectation of a better return on investments. Income generated through tourism, however, often does not directly provide the funds for the conservation of historic buildings. Although tourism has become a key economic sector for many historic towns, tourists visiting a historic town do not pay an entrance fee or money that will directly go to those who have paid for the conservation of the historic buildings either directly or through taxation. Nor will tourists pay the cost of infrastructure supply, although they will make use of it. Nonetheless, tourism often increases awareness of the value of historic buildings and increases investment opportunities. There are also negative impacts of tourism to cultural heritage ranging from damage to historic fabric from large number of visitors to loss of character and poor quality of conservation (see Chapter 6).

The reuse of an existing building is partly about imaginative design solutions but predominantly about economic feasibility. For a project to be successful, the cost of refurbishment and alterations need to be weighed against the economic value of the outcome. As previously discussed, the costs associated with conserving historic properties tend to be higher than those incurred in maintaining relatively new buildings. Nonetheless, creative solutions have demonstrated in places that economic gains can be made where most developers have been cautious to invest. Demand trends can shift and what is

Figure 8.6   This former fish market in Izmir, Turkey, originally built to a design by Gustav Eiffel, has been successfully converted into a popular waterside retail complex.

not viable now, may become so in five years time as a result of other events, investments or demographic and lifestyle shifts.

Covent Garden in London is often held up to be an example of how the listing of the old market turned the area into a popular retail and leisure centre. Today, there is a greater recognition of the potential of historic districts as leisure destinations and even in Shanghai, where wholesale demolition of historic quarters for replacement with mega office complexes is common practice, a few pockets of historic buildings are being retained and developed as a complex of restaurants and bars, creating a unique leisure destination in the city.

Most conservation work is undertaken by public, private or not-for-profit organisations. While until recently these three bodies have remained separate and still are so in some places, increasingly they are collaborating in projects and regeneration initiatives. However, the objectives of the different partners in such collaborations may be very different. For instance, where the public sector is interested in serving the public good, the private partner is more interested in the potential of profit and return on investment.

Especially in places where investors are cautious in taking on a derelict building in a run down area, it may be up to a public sector partner to kick-start revitalisation. For example, in Leeds (England) the use of a nineteenth-century arcade as the centrepiece to city centre shopping and investment in the public realm became the catalyst for the commercial development and investment that followed in the Victoria Quarter. The private sector needs to see a strategy and vision as well as a commitment from the public sector to invest, while the public sector needs the private sector to invest in the area to realise the vision.

Adaptive reuse

Buildings become redundant for a whole host of reasons, from changing economic and industrial practices, demographic shifts, increasing cost of upkeep or maintenance and primarily because they are no longer suited for the function they were being used for and a viable new use has not been identified. For much of their useful life, buildings change incrementally and are continuously updated and adapted to user needs. While the structure remains the most permanent element, changes are likely to be made to the building envelope and more regularly to the internal layout. At the current time, some building services are updated or changed as often as every 7–15 years.

Most buildings have proven to be flexible and with a little adaptation capable of accommodating new uses. Surveys have illustrated how many buildings have undergone several, at times very different, uses since they were first built (see Figures 2.1 and 5.2). This, however, does not mean that all new uses are appropriate for a historic building. If the level of intervention required would cause too much damage to the historic fabric, then this is unlikely to be an appropriate new use. Proposals for a new use must first consider whether the building is appropriate for this use (e.g. does it have enough windows), and secondly, whether the new use and necessary changes protect and enhance the cultural significance of the building. Added to this are the financial considerations linked to demand in the area for such a use and therefore the ability to realise a financially viable project. While a certain building type may prove to be suitable for certain new uses in one area, there may not be a ready market for the same uses in another place. Former industrial buildings may be ideal for loft style apartments in city centre locations, but there will be limited demand for similar developments outside of major urban centres. Like all commercial property, location is also a key defining factor in whether and how a historic building will be conserved and reused.

Conversion and adaptation should enhance the significance and qualities of a historic building. Projects that are often the most successful are those that recognise and work with the 'spirit' of the building. The qualities of a warehouse space, for example, are the large open spaces, the associated flexibility of the plan and the exposed industrial nature of the materials. Dividing up the space into small rooms like those of a semi-detached house will fail to achieve a distinctive space. Furthermore, different building types require different levels of sensitivity when it comes to conversion. The conversion of a church building will need to be more sensitive in choice of a new use, the division of spaces and respect to its former spiritual value.

Figure 8.7 The conversion of the Dreher Brewery in Venice, Italy, into residential units imaginatively makes use of the chimney as the escape staircase.

In recent years, there have been many examples of imaginative new uses for historic buildings. Redundant churches have been converted to arts centres, concert halls and even residential blocks. Victorian water towers and windmills have become family homes, power stations contemporary art galleries and cold stores nightclubs. Sometimes a simple reorganisation or the addition of a new circulation space will make a building viable again (Figure 8.7). It is often the case that the use of an old building will probably give the project a different character from contemporary equivalents. Two recent examples demonstrate very different approaches to reuse. In the case of the conversion of a former power station into Tate Modern in London, a museum of contemporary art, the design benefits from and builds on the vast open space provided by the turbine hall (Figure 8.8). The conversion of the eighteenth-century Royal Naval College buildings, part of the Greenwich World Heritage Site, and designed by some of the great architects of the time including Sir Christopher Wren, Vanbrugh and Hawksmoor, into Greenwich

Figure 8.8 The turbine hall at the Tate Modern, London, a museum of contemporary art converted from a former power station, not only provides a unique space for the display of art works, but has also been an inspiration for artists who have created works specifically for the space.

University was able to create a flagship campus and a new image for the University.

Industrial buildings such as mills were designed for the machinery with which they operated. The conversion of such working buildings can also be difficult, even when the machinery is taken out. As attractive as warehouses and other similar industrial buildings have become for office and residential use, their conversion can also be problematic at times. Built essentially for the safe storage of goods, they are often deep buildings with small openings, while floor to ceiling heights can be limited to the heights originally sufficient for manual labour lifting goods to about head height.

While creative imagination is finding new ways of using old buildings, new buildings and new building types are appearing on the market awaiting a new use. Some buildings, such as a fire station designed by Zaha Hadid for Vitra in Germany and completed in 1994, have had to be adapted to new uses almost as soon as they were built. The fire station has been converted into a museum of chairs. With the end of the Cold War and changes in warfare, more and more military sites are becoming redundant. Many are situated at a considerable distance from major centres of settlement, adding to the difficulty of finding viable new uses for them.

DESIGN IN THE HISTORIC ENVIRONMENT

With few exceptions, the character of cities have changed over time with the architectural styles fashionable in each period as well as the changing demands for their functionality. They have prospered and grown in times of affluence and made do in times of economic depression. New architectural styles and materials are often introduced into the city initially on the periphery and then in empty plots or as replacements for dilapidated buildings. Sometimes, existing buildings are adapted to the new styles. It is very rare that a place has a historic or period unity. Most often, the character is a combination of styles and it is important that this evolution is maintained and the vitality of places can be continued.

Today's architecture is the heritage of tomorrow and another layer in the complex tapestry that makes up the character and identity of cities. One of the objectives of the programme developed for the City of Rotterdam in the Netherlands, to celebrate its designation as 'European Cultural Capital, 2001' was to build 'the future heritage'. New architecture in an existing setting will set out to copy, mimic, integrate, harmonise with, contrast, juxtapose on or even compete with the existing context. Each situation may necessitate a different approach. This section will discuss some of the approaches to design in the historic environment. Above all, it is important to recognise that the most successful architecture is often where the context of the existing environment has been fully understood by the design team.

The context of townscape

An understanding of townscape will inform the conservation and reuse of the existing fabric and new developments and infill within it. Analyses of the townscape are not limited character studies of facades or simply a case of recording typical building materials, styles and detailing. In depth, the understanding of a townscape includes everything from its historic evolution and morphology to the integrity of its various elements. Urban character needs to be seen as a totality, including:

- Urban composition, streets, junctions, the treatment of corners, open spaces, massing, buildings' heights and their relationship to each other;
- Historic evolution and morphology, how has the urban form developed;
- Land use patterns, green spaces and the relationship of different uses to one another;
- Materials used and their regional relevance, how they are used and how they fit together;
- Common typologies relating to components such as openings, roofs and chimneys.

Figure 8.9   The character of these souks in the old town of Jerusalem might be defined as a combination of the architecture, the textures, the play of light and shadow and the deliberately broken vistas, as well as the ongoing imprint of human habitation.

The layout and vistas, sometimes incidental, sometimes considered and others specifically designed, such as model villages or garden suburbs. Furthermore, many monuments, buildings and squares are often designed to be viewed from a certain angle or approach and these perspectives should be respected when alterations are proposed.

Beyond its architectural and town planning attributes, townscape has social, cultural and psychological significance (Figure 8.9). It is an understanding of this asset that is fundamental to successful urban conservation and regeneration projects. In the contemporary city, historic quarters represent a tension between character, aesthetic values and a sense of attachment against economic pressures for development and increases in land value (Figure 8.10). Many historic urban areas are now occupied by people who have no significant past connection to the place, yet there is often a greater sense of attachment to a place of historic character, even when the inhabitants are not associated with that past.

Architectural interventions

New architecture is added to the existing fabric of a city in a number of ways, including:

- New additions or extensions to historic buildings;
- New buildings within gutted interiors or behind retained facades;

Figure 8.10 A historic building continues to exist sandwiched between two larger scale modern buildings in Melbourne, Australia.

- New buildings on empty plots or as replacements for existing buildings (infill);
- Larger developments adjoining historic areas or impacting on their value and character;
- Urban design and townscape improvements.

The immediate considerations for most designers and for planners, who will approve a scheme, will be the external appearance and whether it will mimic the historic or stand out as unequivocally contemporary (Figure 8.11). Neither of these two approaches is necessarily appropriate to each situation and between the two extremes lie a multitude of solutions. Copying the historic, or pastiche, is not authentic or truthful and there is rarely a reason for it, although there are exceptions. On the other hand, a design that is unequivocally of its time will come to be seen as distinctively 'of its time'; the passing of time has not necessarily been kind to many of the bland modernist infill developments of the 1950s and 1960s.

Figure 8.11   A new building in deliberate contrast with a historic building in Melbourne, Australia.

Setting in context

'The conservation of a monument implies preserving a setting which is not out of scale. Wherever the traditional setting exists, it must be kept. No new construction, demolition or modification which would alter the relations of mass and colour must be allowed' (Article 6, The Venice Charter).

New design might involve an extension to a large and significant historic building or a small historic building, a new building in a historic urban environment or a new structure at an archaeological site such as a visitor centre or shelter to protect the ruins. Proposed new additions should be of a quality that will constitute a valuable contribution to the historic building, its setting and the townscape. Alterations and additions should relate to the scale and the proportions of the building and be harmonious with it. Furthermore, the interrelationships that are an integral part of the understanding of the building and its immediate environment should not be compromised (Figure 8.12). In each case, different issues will need to be resolved. An initial consideration must be the impact on the integrity of the whole and the intended setting of the historic building, followed by the technical considerations of a new building

Figure 8.12   The glass roof of a new shop inserted between the Cathedral and its cloisters at Salisbury, England, provides a stunning view of the Cathedral spire.

being placed in close proximity to a historic structure, from foundations to the detailing of joints between the old and new material (Figure 8.14).

In some cases, local or regional planning authorities publish design guides (see also Chapter 4). Design guides vary considerably in content and in direction, in the narrow sense they appear as a style guide. Where they have been thought through they form a robust development framework. Such guides need to clearly differentiate between conservation, additions and new-build projects. They will often provide useful guidelines on building materials typical to the area and therefore of the local character. Many make recommendations for new buildings to use local materials, and in some cases provide a selection of typical local details. This is not necessarily a solution to good architecture. The requirements of a new building are not the same as old buildings and it is important that current needs are also being responded to through appropriate technologies and designs. Furthermore, in some places building materials that were once easily available and affordable may no longer be readily available, especially in the case of timber and stone. The use of artificial variants, however, is rarely a solution to maintaining 'character'.

Any new design should be based on a thorough understanding of the historic building itself and of the townscape or landscape in which it will be situated. This involves not only an analysis of elevation treatments and use of materials, but requires a deeper understanding of their significance and the basis of the original design decisions. When a new building is placed in a street or square there may already exists a collection of buildings of very different styles. It may, therefore, not simply be adjoining buildings from which allegiances are sought. On the contrary, the design process must go beyond this, to look at urban development and how these streets or squares were formed, to understand the design principles/ethos of the place, the greater whole. The scale, massing and height of any new building should not overshadow surrounding historic buildings or detract from the appreciation of established views and open spaces.

With greater demand on land for building, previously empty plots or those that have become empty for various reasons can be prime sites for infill developments. Placing new buildings into spaces that have previously been open spaces, such as many of the extensions to Oxford and Cambridge colleges, will not only introduce new densities but also alter the way spaces around buildings are used and the existing buildings perceived in their setting. Although new volumes into the grain of a historic centre can have more impact than the elevations, this type of development may be necessary to ensure the ongoing vitality of a place.

Smaller interventions into light wells and the like can often solve complex circulation problems as well as accommodate newer needs for lifts and additional staircases. Covering previously open inner courtyards with glass roofs provides the opportunity for otherwise underused or redundant space to become more efficiently used while maintaining light levels into the building (Figure 8.12). In the case of the British Museum in London, the opening up of the courtyard has enabled the introduction of a whole new circulation hub and eased the pressure on other congested areas. While these interventions are not highly visible on the exterior of a historic building or in the townscape, considerations need to be made regarding the impact of the changed environmental conditions on the historic fabric as well structural support of the new roof structure. In some cases, it has been possible to build off existing walls; in other cases, a new structure has been inserted into the courtyard space to support the new roof.

Another approach commonly seen in contextualising new buildings has been the use of glass either for its transparent or reflective qualities. There is an often misguided perception that an elevation clad in glass is transparent and therefore less obstructive in a townscape, yet unintended reflections and glare on the glass may result in a different townscape experience. At the same time, at night when lights are on these buildings can become see-through. In other instances, glass is deliberately used for its reflective properties, in order to fit into the surrounding historic townscape by reflecting it.

Figure 8.13   The extension to the National Gallery in London by Venturi Scott-Brown, where elements of the existing building have been sequentially simplified on the front elevation of the new building.

Figure 8.14   A carefully considered detail between the existing building and new extension at Compton Verney by Stanton Williams Architects.

Figure 8.15 This new development in London's Spitalfields mimics historic styles in an attempt to fit in with the mixed grain of eighteenth century buildings on the other side of the street.

Historicist and pastiche approaches

It was noted above that the morphological growth of a city is the layers of new styles that are introduced in each period. Thus, the introduction of historicist interventions, often in the name of aesthetic conformity, will not necessarily enhance a townscape. One of the arguments put forward in support of rebuilding the demolished royal palace on Berlin's Unter der Linden Avenue (see Chapter 2) was to re-establish the historic integrity of the Avenue.

The principle of minimum intervention should also apply to townscapes, rather than over restoration of facades to perfect frontages. Period-specific authenticity can only be achieved in open air museums, not in the living dynamic of the city. The recreation of a historic elevation or the building of a new building using historic stylistic precedents is often seen as a safe option for infill developments. It is very rare that the design of a pastiche façade has fully considered the historic precedent, its use of materials, their detailing and their proportions. Pastiche buildings often have a tendency of appearing neither historic nor contemporary (Figure 8.15).

Figure 8.16 Architects such as Carlos Scarpa have demonstrated how thoroughly modern materials and details can successfully be incorporated into historic buildings.

Facadism

Facadism is the practice of retaining a historic façade, but constructing a new building behind it. It is often seen as a last resort or compromise solution to maintaining townscape value, while realising the financial potential of a prime site. In some cases, facadism might also refer to the re-construction of a facsimile of the old elevation.

Façade retention, however, is not an equivalent of conservation. The elevation of a building is an expression of the interior and its organisation; elevations are only part of the building, they do not alone represent its integrity. With the loss of a back or interior some of the intrinsic character of the original building will also be permanently lost. While a façade contributes to streetscape character, vistas and views, the building as a whole makes up urban grain, morphology, volume and density. Facadism becomes a way of only looking at external appearances rather than at the totality of an urban area and some more intrinsic qualities, such as community and social values of a place. Changes to the volume and layout of a building not only introduces different uses to an area, but may also bring about morphological changes altering block sizes and density.

Nonetheless, the market favours the historic exterior and the character it adds to a square or street frontage for example, while at the same time demanding or paying a higher premium to the new interior or back. Furthermore, in some instances an elevation might have been specifically designed to contribute a streetscape or square, rather than as an expression of the buildings function. In such an instance, its maintenance is integral to urban conservation.

SUMMARY AND CONCLUSION

Historic buildings can play a vital role in regeneration by recognising the added value of cultural heritage to revitalisation by establishing a unique environment defined by the historic character and sense of local identity that it signifies. Revitalisation can only happen through change and renewal, allowing the historic environment to adapt and integrate with the new. There are numerous ways in which historic buildings can be reused and much can be achieved through the skills of a sensitive and imaginative designer and equally imaginative means of financing projects.

Historic towns and places are continuously changing and evolving to meet the needs of contemporary society. Nonetheless, historic places are valuable assets and any new interventions in the historic fabric of a city must:

- Be based on an understanding of the historic town, its morphological and social development;
- Respect the setting and landscape;
- Be appropriate in scale, height and volume to the inherent morphology of the townscape;
- In design, respect existing characteristics of the townscape and contribute to it rather than mimic or compete with the existing.

FURTHER READING AND SOURCES OF INFORMATION

Cunnington, P. (1988) *Change of Use: The Conversion of Old Buildings*. London, A & C Black Ltd.

Dennison, P. (ed) (1999) *Conservation and Change in Historic Towns*. CBA Research Report. York, Council for British Archaeology.

Eley, P. and Worthington, J. (1984) *Industrial Rehabilitation*. London, Architectural Press.

ICOMOS (1987) *International Charter for the Protection of Historic Towns*.

Larkham, P.J. (1996) *Conservation and the City*. London, Routledge.

Marshall, R. (ed) (2001) *Waterfronts in Post-Industrial Cities*. London, Spon Press.

McKean, J. (2004) *Giancarlo De Carlo: Layered Places*. Stuttgart and London, Editions Axel Menges.

Orbaşli, A. (2000) *Tourists in Historic Towns: Urban Conservation and Heritage Management*. London and New York, Spon Press.

Pickard, R.D. (1996) *Conservation in the Built Environment*. Singapore, Longman.

Richards, J. (1994) *Facadism*. London, Routledge.

Strike, J. (1994) *Architecture in Conservation*. London, Routledge.

Warren, J., Worthington, J. and Taylor, S. (eds) (1998) *Context: New Buildings in Historic Settings*. Oxford, Architectural Press.

Web-based sources

English Historic Towns Forum: www.ehtf.org.uk
European Association of Historic Towns and Regions: www.historic-towns.org
Organisation of World Heritage Cities: www.ovpm.org

Conclusion: Conservation in the Future

The content of this book has taken stock of where conservation is at the start of the twenty-first century. Much of it is based on already published material, opinions expressed by leading conservation professionals and debates that are taking place in a variety of fora. This book has tried to identify just some of the challenges facing conservation professionals today and as the field grows and as participation widens even more challenges lay ahead. The consideration of this 'epilogue' is some of the issues and challenges that will influence the practice of architectural conservation in the coming decades of the twenty-first century. Three immediate concerns lie ahead: climate change, the changing politics of virtual networks and the relevance of cultural heritage to future generations.

Over time the predominantly agrarian economy has been replaced in turn by the industrial economy, the service economy and the knowledge economy. The rate of change is accelerating and present day economies are based on bio-sciences where technology is the driver of change. Shifts in the economy and workplace means that buildings, especially those designed for a specific purpose, are rapidly becoming redundant. Changing workplace practices directly influence how buildings are used or adapted and even how entire areas are regenerated.

In all fields, the alarmingly fast rate of change has serious implications on many traditional management practices, including the management of the historic environment. In cities, urban plans with 5–20 year forecasts may no longer be the relevant means through which to manage a fast evolving environment. Rapid change may be a threat to the historic environment, but it also highlights its unique qualities.

To date, the way industry leaders and large corporations have used property has played an important role in how cities have developed. New technologies are altering the balance of power and the role of location. Wireless networks and the internet have created new social hubs and brought about new ways of working, changing the ways in which the public realm, private and workspaces are used by individuals. Where people have a choice of environment in which to work, character and appeal of the historic environment may well win over the mundane uniformity of a contemporary office. Flexibility and designing for flexibility is increasingly a priority for both design and conservation.

Globalisation has brought with it a cultural uniformity, but it has also played a role in highlighting cultural diversity. Migration, even within the first decade of this century stands at an unprecedented scale. The world over there is an increased mobility of ideas, markets and people. Significant shifts of population will impact on the way cultural heritage is valued by those who are new to a place and those who have left it behind.

In the face of rapid change and technological advancement, the continuous link humankind has had with the past through building practices may well be coming to an end. In many parts of the world, new ways of living, new conveniences and building materials at affordable prices has made traditional ways of building all but redundant. Conservation is often informed by ongoing practices and the loss of know-how and skills brings about an even more urgent need for their preservation. In the same way many facets of the intangible heritage from local foods or costumes to oral traditions are on the brink of disappearing. While this growing gap between traditional and contemporary living increases the pressure to preserve what remains of the vernacular, this is all too often as relic tourism destinations.

At the same time, recognisable monuments and historic themes adorn theme parks and Las Vegas. At a time where the conservation of historic urban quarters is a hard-won battle, they re-emerge as the streetscapes of theme parks and holiday resorts. Seen as a major economic contributor, tourism can play a key role in how the cultural heritage is used, conserved and presented. Despite fears of climate change, travel, tourism and the leisure sector are likely to continue their growth and to nurture the ongoing commoditisation of cultural heritage. How soon virtual reality, as a tool through which the historic environment can be experienced and interpreted, will take over from the real thing and an 'authentic' experience is yet to be seen.

The full impact of climate change is still to be fully understood, but the maintenance, conservation and management of historic buildings will undoubtedly be affected. Historic buildings will need to be adapted to cope with changing climatic conditions, flooding, increased wind loads, heavy storms and rainfall. Choices for sustainable buildings are increasingly considering the embodied energy of production and transportation costs, amongst other 'green' criteria. In this respect, the reuse of existing buildings makes environmental sense. Nonetheless, many historic buildings fall short of the energy performance targets that are now being set. With pressures to reduce carbon emissions and predicted rises in fuel prices, the large stock of existing buildings that are not performing as efficiently as new buildings need to be considered. This is partly due to the way in which they were used historically and the change in comfort expectations of users in terms of both heating and cooling. Research into ways in which these buildings can be upgraded is essential at this time.

In terms of innovation, the construction industry continues to be a follower rather than a leader. Nonetheless, some recent innovations do have implications for architectural conservation, from investigative technologies that are developing, to GIS systems and integrated databases that assist in the more effective management of the cultural heritage. Meanwhile, communications and networking technologies have significantly altered how professionals communicate and how information is accessed and shared, and will continue to do so. However, while some of these technological advancements create greater equality and access to information globally, they will also further emphasise the gap between developed and developing countries.

As the scope of cultural heritage is widening, so is the role of the conservation professional and new skills that may need to be brought to the team. The conservation and cultural resource management skills that have been passed down so far may not be the right tools with which to approach the new dimensions of heritage and the new ways of working. The relevance of conservation will also, more than ever, have to continue to be justified, either in economic terms or through other measurable targets.

The validity of historic buildings and places are often seen in terms of the value they present to those who use or interact with them. Yet, the cultural heritage is an asset that has been passed down to the present generation and one which must be passed onto others after us, not simply consumed by the present for whatever purpose. The role of the conservation professional remains as a caretaker, maintaining the asset for future generations, while facilitating the changes that will make this possible.

Glossary

adaptive reuse
Making changes to accommodate a new use to continue the usefulness of a historic building.

arris
Sharp edge created at the junction of two surfaces, e.g. of two sides of a brick.

authenticity
Being genuine or original. In the case of cultural heritage may refer to building materials, design or decorative concepts, location and setting.

breathability
Ability of the building to breathe by allowing moisture to flow through all layers of the building fabric.

climate change
Changes to the climate of a region or the Earth as a whole, an outcome of global warming caused by human activity.

compressive strength
Ability of a material to withstand compressive forces in the axial direction. When the limit is reached crushing occurs.

conservation
'All the processes of looking after a place so as to retain its cultural significance' (Burra Charter, 1999).

conservation management plan
A framework policy document that will inform the conservation and management of a building or place of cultural significance.

consolidation
Physical interventions undertaken to stop further decay and structural instability.

cultural heritage
Tangible relics from the past and the intangible values associated with them that are of cultural and/or historic significance to a society.

cultural landscape
A geographic area that is formed or influenced by humankind, seen as the totality of its built, cultural and natural attributes.

decay
Breakdown of organic or chemical components of a material.

defect
A failing or shortcoming in the performance of a building's materials, structure or services.

degradation
Micro-scale or small-scale deterioration.

deterioration
Damage and subsequent change in a material caused by a range of climatic, biological, natural and human factors.

distress
The process through which materials are worn down by use and natural factors.

integrity
Wholeness, in reference to the historic, architectural or artistic integrity of a building.

intumescent
A substance that will swell when exposed to heat, commonly used in fire protection.

permeability
Ability of a material to transmit fluids.

preservation
To maintain a building in its existing form and condition. Often used in the US in the same way as conservation is used in the UK.

prevention/preventative measures
To alter conditions so as to reduce or slow the process of decay or deterioration.

reconstitution
Building back a collapsed building or parts of it piece by piece.

reconstruction
Re-creation by building a replica of a building on its original site.

regeneration
Economic and environmental development with the objective of improving physical and socio-economic conditions in order to realise a more vibrant and dynamic environment.

repair
Necessary interventions to make good a building or its fabric.

replica
A facsimile copy of a historic building or part of it.

restoration
Returning a building or parts of it to a form in which it appeared at some point in the past.

risk
Potential negative impact to the significance or value of the cultural heritage.

stabilisation
Strengthening of a building structure or materials in situ to improve its structural stability.

stability
Ability of a structure to stand up as a whole.

stiffness
Ability of a building structure to support loads without undesirable movement.

strength
The ability of a building structure or its components to carry dead (its own) and applied loads.

structural strengthening
Increasing or restoring structural load bearing capacity of a structure.

sustainable development
'Development that meets the needs of the present generation without compromising the ability of future generations to meet their own needs' (Bruntland Report, 1987).

tensile strength
Ability of a material to withstand tensile stresses before it is permanently deformed or fails.

townscape
The configuration of urban forms, architectural features, open spaces and green areas that make up the visual character of a town.

urban conservation
The process that seeks economies that will facilitate the conservation and sustainability of the historic environment.

u-value
Calculation of the thermal resistance of each layer of a building's fabric as heat passes from inside to outside.

wear and tear
Wearing of the building fabric as a result of normal daily usage.

Bibliography

Appleyard, D. (ed) (1979) *The Conservation of European Cities*. Cambridge, MA, MIT Press.

Ashurst, J. and Ashurst, N. (1988) *Practical Building Conservation, Vol. 3: Mortars, Plasters and Renders*. Aldershot, Gower Technical Press.

Ashurst, J. and Ashurst, N. (1988) *Practical Building Conservation, Vol. 5: Wood, Glass & Resin*. Aldershot, Gower Technical Press.

Ashurst, J. and Ashurst, N. (1990) *Practical Building Conservation, Vol. 1: Stone Masonry*. Aldershot, Gower Technical Press.

Ashurst, J. and Ashurst, N. (1991) *Practical Building Conservation, Vol. 4: Metals*. Aldershot, Gower Technical Press.

Ashurst, J. and Ashurst, N. (1995) *Practical Building Conservation, Vol. 2: Brick, Terracotta and Earth*. Aldershot, Gower Technical Press.

Ashurst, J. and Dimes, F.G. (eds) (1998) *Conservation of Building and Decorative Stone*, Paperback edn. Oxford, Butterworth-Heinemann.

Australia ICOMOS (1999) *The Burra Charter*.

Baer, N.S. and Snethlage, R. (eds) (1997) *Saving Our Architectural Heritage: The Conservation of Historic Stone Structures*. Chichester, John Wiley & Sons.

Binney, M. (1985) *Our Vanishing Heritage*. London, Arlington Books.

Binney, M. and Hanna, M. (1978) *Preservation Pays: Tourism and the Economic Benefits of Conserving Historic Buildings*. London, Save Britain's Heritage.

Brand, S. (1994) *How Buildings Learn*. New York, Viking Penguin.

Bristow, I.C. (1986) *Architectural Colour in British Interiors 1615-1840*. London, New Haven, Yale University Press.

Brown, A. (1998) *Windsor Castle Fire and Restoration*. Windsor, Cobblestone Communications.

Brunskill, R.W. (1992) *Traditional Buildings of Britain: An Introduction to Vernacular Architecture*. London, Victor Gollancz Ltd.

Burman, P., Pickard, R. and Taylor, S. (eds) (1995) *The Economics of Architectural Conservation*. York, Institute of Advanced Architectural Studies.

Burman, P. and Stratton, M. (eds) (1997) *Conserving the Railway Heritage*. London and New York, Spon Press.

Cantacuzino (1989) *Re-architecture: Old Buildings/New Uses*. New York, Abbeville Press.

Caroe, A. and Caroe, M. (2001) *Stonework: Maintenance and Surface Repair*, 2nd edn. London, Church House.

Chanter, B. and Swallow, P. (1996) *Building Maintenance and Management*. Oxford, Blackwell Science.

Charles, F.W.B. and Charles, M. (1984) *Conservation of Timber Buildings*. Cheltenham, Stanley Thomas Ltd.

Clark, K. (2001) *Informed Conservation*. London, English Heritage.

Cox, J. and Letts, J. (2000) *Thatch: Thatching in England 1940–1994 English Heritage Research Transactions*, Vol. 6. London, James and James.

Croci, G. (1998) *The Conservation and Structural Restoration of Architectural Heritage*. Southampton, Computational Mechanics Publications.

Cunningham, A. (ed) (1998) *Modern Movement Heritage*. London, Spon Press.

Cunnington, P. (1988) *Change of Use: The Conversion of Old Buildings*. London, A & C Black Ltd.

David, J. (2002) *Guide to Building Services for Historic Buildings*. London, CIBSE.

Delafons, J. (1997) *Politics and Preservation*. London, Spon Press.

Dennison, P. (ed) (1999) *Conservation and Change in Historic Towns*. CBA Research Report. York, Council for British Archaeology.

Denslagen, W. (1994) *Architectural Restoration in Western Europe: Controversy and Continuity*. Amsterdam, Architectura and Natura Press.

Earl, J. (1996) *Building Conservation Philosophy*. Reading, College of Estate Management.

Eley, P. and Worthington, J. (1984) *Industrial Rehabilitation*. London, Architectural Press.

English Heritage (2000) *Power of Place*. London, English Heritage.

English Historic Towns Forum (1998) *Conservation Area Management: A Practical Guide*. Report No. 38, English Historic Towns Forum.

Erder, C. (1986) *Our Architectural Heritage: From Consciousness to Conservation*. Paris, UNESCO.

Feilden, Sir B.M. (2003) *Conservation of Historic Buildings*, 3rd edn. Oxford, Architectural Press.

Fitch, J.M. (1992) *Historic Preservation: Curatorial Management of the Built World*. Charlotetsville, University Press of Virginia.

Foster, L. (1997) *Access to the Historic Environment: Meeting the Needs of Disabled People*. Shaftesbury, Donhead.

Gause, J.A. (1996) *New Uses for Obsolete Buildings*. Urban Land Institute, US.

Glasson, J. *et al.* (eds) (1999) *Introduction to Environmental Impact Assessment*. London and New York, Spon Press.

Hewison, R. (1986) *The Heritage Industry*. London, Methuen.

Holmes S. and Wingate, M. (1997) *Building with Lime: A Practical Introduction*. London, Intermediate Technology Publications.

Highfield, D. (2000) *Refurbishment and Upgrading of Buildings*. London, Spon Press.

Hughes, H. (ed) (2002) *Layers of Understanding: Setting Standards for Architectural Paint Research*. Dorset, Donhead Publishing.

Hunter, M. (ed) (1996) *Preserving the Past: The Rise of Heritage in Modern Britain*. Gloucester, Great Britain, Alan Sutton Publishing.

ICOMOS (1964) *The Venice Charter*.

ICOMOS (1984) *ICOMOS 1964–1984*. Paris, ICOMOS.

ICOMOS (1987) *International Charter for the Protection of Historic Towns*.

ICOMOS (1993) *Guidelines on Education and Training in the Conservation of Monuments, Ensembles and Sites*.

Institution of Civil Engineers (1989) *Conservation of Engineering Structures*. London, Thomas Telford.

Jokilehto, J. (1999) *A History of Conservation*. Oxford, Butterworth Heinemann.

Kidd, S. (ed) (1995) *Heritage Under Fire*. London, Fire Protection Association for the UK Working Party on Fire Safety in Historic Buildings.

Larkham, P.J. (1996) *Conservation and the City*. London, Routledge.

Larsen, K.E. and Marstein, N. (2000) *Conservation of Historic Timber Structures*. Oxford, Butterworth Heinemann.

Latham, D. (2000) *Creative Reuse of Buildings*, Vols. 1 and 2. Dorset, Donhead.

Levy, M. and Salvadori, M. (1992) *Why Buildings Fall Down: How Structures Fail*. New York, W.W. Norton.

Lichfield, N. (1988) *Economics in Urban Conservation*. Cambridge, Cambridge University Press.

Lynch, G.C.J. (1990) *Gauged Brickwork: A Technical Handbook*. Aldershot, Gower Technical.

Lynch, G.C.J. (1994a) *Brickwork: History, Technology and Practice*, Vol. 1. London, Donhead Publishing.

Lynch, G.C.J. (1994b) *Brickwork: History, Technology and Practice*, Vol. 2. London, Donhead Publishing.

Macdonald, S. (ed) (1996) *Modern Matters: Principles and Practice in Conserving Recent Architecture*. Dorset, Donhead.

Marshall, R. (ed) (2001) *Waterfronts in Post-Industrial Cities*. London, Spon Press.

McKean, J. (2004) *Giancarlo De Carlo: Layered Places*. Stuttgart and London, Editions Axel Menges.

Miele, C. (ed) (2005) *William Morris: Building Conservation and the Arts and Crafts Cult of Authenticity, 1877–1939*. New Haven and London, Yale University Press.

Moir, J. and Letts, J. (1999) *Thatch: Thatching in England 1790–1940 English Heritage Research Transactions*, Vol. 5. London, James and James.

Morris, P. and Therivel, R. (eds) (2001) *Methods of Environmental Impact Assessment*, 2nd edn. London and New York, Spon Press.

Oliver, P. (ed) (1997) *Encyclopedia of Vernacular Architecture of the World. Vol. 1: Theories and Principles*. Cambridge, Cambridge University Press.

Orbaşli, A. (2000) *Tourists in Historic Towns: Urban Conservation and Heritage Management*. London and New York, Spon Press.

Oxley, R. (2003) *Survey and Repair of Traditional Buildings: A Sustainable Approach*. Shaftesbury, Donhead.

Pavia, S. and Bolton, J. (2000) *Stone, Brick and Mortar: Historical Use, Decay and Conservation of Building Materials in Ireland*. Wordwell, Bray, Co. Wicklow.

Pearce, D. (1989) *Conservation Today*. London, Routledge.

Phelps, A., Ashworth, G.J. and Johansson, B.O.H. (eds) (2002) *The Construction of the Built Heritage*. Aldershot, Great Britain, Ashgate.

Phillips, D. (1997) *Lighting Historic Buildings*. Oxford, Architectural Press.

Pickard, R.D. (1996) *Conservation in the Built Environment*. Singapore, Longman.

Pickard, R.D. (ed) (2001) *Management of Historic Centres*. London, Spon Press.

Pickard, R. (ed) (2001) *Policy and Law in Heritage Conservation*. London, Spon Press.

Rabun, J.S. (2000) *Structural Analysis of Historic Buildings: Restoration, Preservation and Adaptive Reuse Applications for Architects and Engineers*. Chichester, John Wiley & Sons.

Richards, J. (1994) *Facadism*. London, Routledge.

Richards, R. and Urquhart, M. (2003) *Conservation Planning*, 2nd edn. London, Planning Aid for London Publications.

Roberts, P. and Sykes, H. (2000) *Urban Regeneration: A Handbook*. London, Sage.

Robson, P. (1991) *Structural Appraisal of Traditional Buildings*. Dorset, Donhead.

Robson, P. (1999) *Structural Repair of Traditional Buildings*. Dorset, Donhead.

Ross, P. (2002) *Appraisal and Repair of Timber Structures*. London, Thomas Telford.

Rowell, C. and Robinson J.M. (1996) *Uppark Restored*. London, National Trust.

Royal Town Planning Institute (2000) *Conservation of the Historic Environment: Good Practice Guide*. London, RTPI.

Salvadori, M. (1990) *Why Buildings Stand Up: The Strength of Architecture*. New York, W.W. Norton.

Sample Kerr, J. (1996) *The Conservation Plan*, 4th edn. Sydney, National Trust of Australia (NSW).

Stratton, M. (ed) (1997) *Structure and Style: Conserving 20th Century Buildings*. London, E & FN Spon.

Stratton, M. (ed) (2000) *Industrial Buildings: Conservation and Regeneration*. London, E & FN Spon.

Strike, J. (1994) *Architecture in Conservation*. London, Routledge.

Suddards, R.W. and Hargreaves, J.M. (1996) *Listed Buildings*, 3rd edn. Gloucester, Sweet & Maxwell Limited.

Swallow, P. *et al.* (2004) *Measurement and Recording of Historic Buildings*, 2nd edn. Shaftesbury, Dorset, Donhead.

Teutonico, J.M. and Palumbo, G. (eds) (2000) *Management Planning for Archaeological Sites*. Los Angeles, CA, J. Paul Getty Trust.

The Aga Khan Trust for Culture (1990) *Architectural and Urban Conservation in the Islamic World*. Geneva, Switzerland, The Aga Khan Trust for Culture.

Tiesdell, S., Oc, T. and Heath, T. (1996) *Revitalising Historic Urban Quarters*. Oxford, Butterworth-Heinemann.

Timothy, D.J. and Boyd, S.W. (2003) *Heritage Tourism*. London, Prentice Hall.

TRADA Technology Ltd. (2001) *Timber Frame Construction*, 3rd edn. High Wycombe, TRADA Technology.

UNESCO (1994) *The Nara Document on Authenticity*.

UNESCO (2005) *Operational Guidelines for the Implementation of the World Heritage Convention*. Paris, World Heritage Centre. Available at www.unesco.org/culture

von Droste, B., Plachter, H. and Rössler, M. (eds) (1995) *Cultural Landscapes of Universal Value*. Stuttgart and New York, Gustav Fischer Verlag.

Walker, R. (1995) *The Cambridgeshire Guide to Historic Buildings Law*. Cambridge, Cambridgeshire County Council.

Warren, J. (1998) *Conservation of Brick*. Oxford, Butterworth-Heinemann.

Warren, J. (1999) *Conservation of Earth Structures*. Oxford, Butterworth-Heinemann.

Warren, J., Worthington, J. and Taylor, S. (eds) (1998) *Context: New Buildings in Historic Settings*. Oxford, Architectural Press.

Watt, D.S. (1999) *Building Pathology: Principles & Practice*. Oxford, Blackwell Science.

Watt, D. and Colston, B. (eds) (2003) *Conservation of Historic Buildings and Their Contents*. Dorset, Donhead.

Watt, D. and Swallow, P. (1996) *Surveying Historic Buildings*. Shaftesbury, Dorset, Donhead.

Weaver, M.E. (1997) *Conserving Buildings: A Manual of Techniques and Materials.* New York, John Wiley & Sons.

Yeomans, D. (1992) *The Trussed Roof: Its History and Development.* Aldershot, Scolar Press.

– (1979) *Ancient Monuments and Archaeological Areas Act.* London, HMSO.

– (1990) *Planning (Listed Buildings and Conservation Areas) Act.* London, HMSO.

– (1990) *Town and Country Planning Act.* London, HMSO.

– (1990) *Planning Policy Guidance 16: Archaeology and Planning.* London, HMSO.

– (1994) *Planning Policy Guidance 15: Planning and the Historic Environment.* London, HMSO.

– (2000) *Terra 2000 Preprints: 8th International Conference on the Study and Conservation of Earthen Architecture,* Torquay, Devon, UK, May 2000. London, James & James.

– (2000) *Terra 2000 Postprints: 8th International Conference on the Study and Conservation of Earthen Architecture,* Torquay, Devon, UK, May 2000. London, James & James.

Journals and periodicals

Ancient Monuments Society Transactions

Architectural Heritage, Journal of the Architectural Heritage Society of Scotland

Association for Studies in the Conservation of Historic Buildings Transactions

Building Conservation Journal, Royal Institution of Chartered Surveyors

Context, Journal of the Institute of Historic Building Conservation

English Heritage Guidance Notes

English Heritage Research Transactions

Getty Conservation Newsletter: also available online from www.getty.edu/conservation/publications/newsletters/index.html

Historic Environment, Australia ICOMOS Journal

International Journal of Architectural Heritage, Taylor and Francis

Journal of the American Institute of Conservation

Journal of Architectural Conservation, Donhead

Journal of Urban Design, Taylor and Francis

Preservation, the Magazine of the National Trust for Historic Preservation, USA

Studies in the History of Gardens & Designed Landscapes, Taylor and Francis (formerly the *Journal of Garden History*)

The Georgian Group Journal

Traditional Dwellings and Settlements Review, University of California at Berkeley

Twentieth Century Architecture, Journal of the Twentieth Century Society

Vernacular Architecture, Journal of the Vernacular Architecture Group, Maney Publishing

Victorian, Magazine of the Victorian Society

Web-based sources

Conservation Information Network (Canada but with international links): www.bcin.ca

Conservation On-Line, Stanford University, USA: http://palimpsest.stanford.edu

Cornell University, USA, database and internet resource for conservation professionals: www.preservenet.cornell.edu

Council of Europe: www.coe.int

English Heritage: www.english-heritage.org.uk
English Historic Towns Forum: www.ehtf.org.uk
Europa Nostra: www.europanostra.org
European Association of Historic Towns and Regions: www.historic-towns.org
Getty Conservation Centre: www.getty.edu/conservation
Getty Conservation Newsletter: www.getty.edu/conservation/publications/newsletters/ index.html
Heritage Lottery Fund, UK: www.hlf.org.uk
Historic Scotland: www.historic-scotland.gov.uk
ICOM conservation committee (materials conservation): www.icom-cc.org
Institute of Historic Building Conservation: www.ihbc.org.uk
International Centre for the study of the Preservation and Respration of Cultural Property (ICCROM): www.iccrom.org
International Council on Monuments and Sites (links to national committees and international scientific committees): www.icomos.org
International Council of Museums (ICOM): www.icom.org
International Union of Nature Conservation: www.iucn.org
National Trust, England: www.nationaltrust.org.uk
National Trust of Australia: www.nattrust.com.au
New Zealand Historic Places Trust: www.historic.org.nz
Organisation of World Heritage Cities: www.ovpm.org
Royal Commission on the Ancient and Historical Monuments of Scotland: www. rcahms.org.uk
Royal Commission on the Ancient and Historical Monuments of Wales: www. rcahmw.org.uk
Royal Institute of British Architects and British Architectural Library: www. architecture.com
Save Britain's Heritage (link to Save Europe's Heritage): www.savebritainsheritage.org
Society for the Protection of Ancient Buildings: www.spab.org.uk
The Aga Khan Trust for Culture: www.akdn.org/agency/aktc.html
The Civic Trust, England: www.civictrust.org.uk
The Georgian Group: www.georgiangroup.org.uk
The Scottish Civic Trust: www.scottishcivictrust.org.uk
The World Bank: www.worldbank.org
UK Government National Archives: www.nationalarchives.gov.uk
UNESCO conventions, projects, World Heritage Sites, Intangible Heritage information: www.unesco.org/culture
United States National Park Service: www.ups.gov
USA: www.eculturalresources.com
Victorian Society: www.victoriansociety.org.uk
Victorian Society in America: www.victoriansociety.org
Welsh Historic Monuments, Cadw: www.cadw.wales.gov.uk
World Monument Fund: www.wmf.org
    www.buildingconservation.com

Although all these websites were current at time of going to print, websites and their contents are liable to change.

Index